LOW TECH EDUCATION IN A HIGH TECH WORLD

LOW TECH EDUCATION IN A HIGH TECH WORLD

Corporations and Classrooms
in the New Information Society

Elizabeth L. Useem

Issues in Science and Technology Series
American Association for the Advancement of Science

THE FREE PRESS
A Division of Macmillan, Inc.
NEW YORK

Collier Macmillan Publishers
LONDON

Copyright © 1986 by American Association for the Advancement
of Science, Inc.

The Free Press
A Division of Macmillan, Inc.
866 Third Avenue, New York, N.Y. 10022

Collier Macmillan Canada, Inc.

Printed in the United States of America

printing number

1 2 3 4 5 6 7 8 9 10

Library of Congress Cataloging-in-Publication Data

Useem, Elizabeth L.
Low tech education in a high tech world.

(AAAS issues in science and technology series)
Bibliography: p.
Includes index.
1. Industry and education—California—Santa Clara
County—Case studies. 2. Microelectronics industry—
Social aspects—California—Santa Clara County—Case
studies. 3. Industry and education—Massachusetts—
Boston—Case studies. 4. Microelectronics industry—
Social aspects—Massachusetts—Boston—Case studies.

I. Title. II. Series.
LC1085.U84 1986 370.19'7 85-16044
ISBN 0-02-933150-1

For Jerry, Andrea, and Susan

Contents

Acknowledgments

This research effort was made possible by the support of many people. My work in Silicon Valley was facilitated by a visiting faculty appointment at the University of California at Santa Cruz during 1980–81, a position made possible by John Kitsuse and Dane Archer in the sociology department there. The research on Route 128 in the Boston area in 1981–82 was supported by a Faculty Fellowship from the National Institute of Education, which was administered by the Institute for the Interdisciplinary Study of Education (IISE) at Northeastern University. The director of IISE, Pat Golden, and the associate directors, Irene Nichols and Holly Carter, were especially helpful throughout that year of research. Staff members of IISE, especially Tracey Boyce, Michael Bernard, Mary Greeley, and Paula Amir, aided me in countless ways. Several Northeastern administrators, including Karl Weiss, Ray Williams, and Paula Leventman, gave me invaluable assistance. Paul Attewell at the State University of New York at Stoneybrook was a source of ideas throughout the project. Kevin Mulvey at the University of Massachusetts at Boston provided additional help on research details. Jerry Useem gave technical computer assistance in the preparation of the manuscript. I have benefited also from the institutional support of the University of Massachusetts at Boston.

I am exceedingly grateful to the 230 individuals in industry, education, and government who agreed to be interviewed for this study. They were generous with their time and candid in the assessments they shared with me. A number of them were interviewed more than once. I am also indebted to the mathematics and science teachers and department chairs in the following Massachusetts high schools: Burlington, Bedford, Waltham, Newton North, Newton South, Woburn, Billerica, and Lexington. Their cooperation in completing my survey questionnaire was indispensable. Dale Carlson at the California State Department of Education was also helpful in providing me with research studies conducted by the California Assessment Program. And Lenny Siegel at the Pacific Studies center in Mountain View, California, generously allowed me to use the Center's files. The late Joseph Bellenger of Los Gatos, California, shared ideas and findings with me as we pursued companion studies.

Finally, my greatest debts are to Paul Osterman and Michael Useem, who gave me continuous intellectual and moral support from the start of this project to its conclusion.

ELIZABETH L. USEEM

1 Technological Boom, Educational Gloom

As America enters the microelectronics era, its educational system is in decline. The affluence and dynamism of high technology companies contrast starkly with the pauperism and deterioration of public educational institutions. Beset by fiscal austerity, decreasing enrollments, and waning standards of student performance, today's schools seem less capable in meeting the needs of our increasingly technically oriented economy than ever before. This development is most graphically illustrated in the two leading centers of high technology in the United States—northern California's Santa Clara County ("Silicon Valley") and the area around Route 128 ("America's Technology Highway") in suburban Boston—where educators are preoccupied with managing decline while neighboring corporate executives plan innovative products for growing markets. The bitter irony of educational despair in a time of industrial boom presents our own decade with a historic challenge.

School directions and industrial trends have never been perfectly synchronized. Employers have long complained that entry-level workers lack many of the skills and attitudes needed for competent job performance, and that vocational programs have trained students for nonexistent jobs. Moreover, the relationship between com-

money

panies and educational institutions, which has waxed and waned over the course of U.S. history, has always been problematic. In some eras, the influence business has exerted over schools and colleges has been too strong and too self-interested, while at other times it has assigned school needs a low priority in its spectrum of concerns. During the first industrial revolution, private corporations played a direct role in shaping the goals and organization of American public schools, sometimes warping the institutions to meet their own narrow goals of productivity and profit. However, in the years prior to the launching of Sputnik, firms were relatively indifferent to the inadequacies of American schooling.

Today, the new industrial revolution created by microelectronic devices forces us to reevaluate connections between education and the private sector. The disparity between the capacities and goals of education and industry has widened and threatens the welfare of workers, companies, and the U.S. economy as a whole. While there are legitimate fears that companies may become too influential in certain aspects of education, it is corporate indifference to education today rather than interference that poses the greater danger. True, there has recently been a concern for a revival of education among employers, evidenced by the growth in the number of collaborative programs between firms and educational institutions. Business has become more active in school politics at the state level and in vocational and job training. Company contributions of equipment have made a substantial difference in the development of some new high technology curricular offerings, and urban partnership programs have helped boost the morale of inner-city school systems. Yet the need for corporate support of education, especially in technology-oriented fields, far outstrips the help currently being supplied.

Present

urbl

Despite the "tidal wave" of reform legislation passed in the mid-1980s, public schools, colleges, and universities do not have adequate resources to provide students with the kind of education they need to prosper in an increasingly technologized labor market. Sorely needed help from the federal government is not in sight. State governments, which have recently moved energetically to require higher academic standards and to provide more money for education, will not be able to sustain the long-term funding increases that are needed to revitalize the educational system. While the nation's corporations have shown renewed concern for the plight of the schools, their support for change often falters when hefty price tags are attached to publically financed reform measures.

The gap between corporate directions and educational conditions is especially apparent in the areas of science, mathematics, and technical instruction. The new industrial age is driven by sophisticated microelectronics, yet training in the disciplines that underlie the technology has eroded. Engineering and other forms of technical education suffer from outdated equipment and an insufficiency of qualified instructors. Secondary science and mathematics education have been weakened by budget restrictions and reductions, teacher shortages, and the general loosening of curriculum standards. In Silicon Valley and the Route 128 areas, as in the rest of the nation, science teachers try to cope with inadequate supplies in ill-equipped laboratories, and engineering professorships remain unfilled for lack of qualified applicants. Despite the extensive publicity these conditions received during the early 1980s, little has been done to alleviate the problem. The situation continues to deteriorate.

Business executives, as a result, are justifiably concerned that shortages of scientific and technically trained personnel will inhibit the growth of their companies and will give foreign competitors with better-trained work forces a significant advantage in product development. Employers in all occupational sectors claim that many of the graduates of the nation's schools and colleges have significant deficiencies in communications and critical thinking skills, and in their knowledge of mathematics and science. National government leaders are concerned about the international competitive position of whole industries while state and local public officials fear that other regions in the country with superior educational systems will attract and succeed in winning technologically oriented firms.

It is not only the health of particular companies and industries that is imperiled by the deficiencies in our educational system, but the individual worker's well-being as well. The number of unskilled jobs in basic industries is diminishing as a result of recession, international competition, automation, and the flight of unskilled assembly work to nations with a cheaper labor force. Diverse groups of workers find the nature of their tasks changing as new technologies alter the workplace. Retraining and reemploying dislocated workers is especially difficult when they lack a solid education in basic reading, writing, and mathematics skills. Those most likely to seek and benefit from retraining programs are people whose academic foundation is already strong.[1] The ability to communicate well and to approach issues analytically—skills developed in liberal arts courses—is of increasing importance in the workplace. These facts underscore the need for an academically sound, broad-based education in order to acquire more highly skilled jobs and to adapt to periodic changes in employment conditions.

A higher level of scientific literacy is also essential for adults in their roles as consumers and citizens. The number and complexity of political and personal decisions requiring at least some rudimentary scientific understanding are increasing, a phenomenon that compels schools to upgrade student learning. For example, people need a scientific understanding of issues affecting nutrition and health, occupational safety, environmental pollution, and the nuclear arms race. Some facility with numbers and statistics enables voters to assess public fiscal questions intelligently. A certain amount of technical proficiency protects consumers as they purchase or repair complicated products.

These issues are all tied to the broader question of how closely educational institutions respond to changes in the economy. In certain fundamental ways, American schools since their inception have adapted their goals, organization, and programs to fit industrial requirements. Manufacturing firms in the early twentieth century, for example, were able to reorganize the public schools along lines more congruent with their needs for a regimented, stratified, and reliable work force. Corporate influence (often wielded through the National Association of Manufacturers) was an important force in the development of vocational education, I.Q. testing, classification of students into ability groups, and modern methods of "scientific management" in schools.[2] Many of these practices remain standard operating procedure in schools today.

But schools today are subject to influences outside the economic order as well. They respond to internal organizational demands and to the needs of a wide number of interest groups in society.[3] The proliferation of groups that influence educational practices and the fragmented nature of school governance have led to a situation where "no one is in charge."[4] Schools are no longer responsive to any one group, not even business. If one examines the evidence of educational decline in the United States in the last two decades, it is difficult to prove that schools and institutions of higher education are successfully meeting the work force requirements of industry. While historical materials from the first industrial revolution in the United States show that leading manufacturers played an important role in shaping educational policy, the process of educational change unfolding in the "new information society" may not be developing along similar lines.

The two high technology "hot spots" in the United States—Silicon Valley in California and Boston's Route 128 corridor—provide an excellent laboratory for examining the relationship between economic directions and educational practices within the context of the "second industrial revolution." In these two regions, manufacturing is dominated by computers and electronics, and both areas promise to remain the most important sites of innovation in technologically sophisticated products for the foreseeable future. These two areas of the country are both a microcosm and a harbinger of the way manufacturing environments will look in future decades. The trade associations representing companies in both regions have issued many reports publicizing their need for a more ample supply of educated workers in recent years. Thus, educational trends in the Boston and Santa Clara Valley areas can illuminate the issue of whether or not educational institutions are being reoriented to develop the technical and communication skills essential for the electronics manufacturing and service economy.

The Decline and Reassessment of Technical and Scientific Education

Computer background

In an era where a more rigorous academic education is needed by a larger proportion of students, the quality of American schooling has waned. Although the United States succeeds in educating a higher percentage of children for a longer period of time and at higher per capita expenditure levels compared to other countries, the public schools experienced a decline in achievement levels from the early 1960s to the early 1980s, a drop in enrollment after the early 1970s, and a reduction in teacher quality and funding levels beginning in the late 1970s.[5] The declines have sparked a host of national studies, published in 1983 and 1984, written by the educational research community and by task forces of government, business, and educational administrators.[6] Since the deterioration of American schools has become apparent at a time of a stagnating manufacturing economy, stepped-up international competition, and a changing industrial base, a special urgency has characterized the debate about the need for educational change. Since continued leadership in microelectronic and other advanced technologies will be essential if the United States is to retain a strong economy, the nation's schools must turn out an educated work force competent to meet this task.

At a time when school administrators have needed to focus attention on the academic curriculum, they instead have been preoccupied with the management of enrollment declines and reduced budgets. The massive drop in the number of students as a result of falling birthrates has brought wrenching cutbacks and upheaval to American education since the early 1970s. There was a 15 percent drop in elementary school enrollment between 1971 and 1981 and a drop of 7 percent in secondary school enrollment (which peaked in 1976–77) during that same period. Elementary schools lost an average of nearly 600,000 students a year during the 1970s. More than 3,600 school buildings were closed between 1971 and 1981, most of them elementary schools. Secondary school enrollment will fall an additional 16 percent between 1980 and 1990. High school and junior high school closings and an increase in the grade span of these schools will become more frequent throughout the 1980s as the full impact of the decline in births is felt in the upper grades. The number of college mergers and closings will also increase. After 1985, the school-age population will begin to rise with an 18 percent overall increase projected between 1985 and the year 2000.[7] The ferment surrounding the sizable teacher layoffs and school closings caused by declining enrollments has absorbed the attention of administrators and school boards, robbing them of time needed to study and change educational curricula and school organization.

Rather than implementing changes that would upgrade science, mathematics, and technical education in these early years of the microelectronics era, reductions in budgets have forced educators to reduce the quality of instruction in those fields. Between 1978 and 1981, public school expenditures per pupil (adjusted for inflation) actually reversed the annual rise that has characterized educational budgets since 1940. Spending has increased modestly in real dollars since then.[8] But expenses for books, materials, and fuel have inflated faster than the consumer price index in the last decade, and schools have also been required to provide costly new programs for pupils with learning disabilities as well as for handicapped and bilingual students. Between the mid-1960s and 1983, there was a 50 percent reduction in expenditures for textbooks and other educational materials.[9] Science equipment and materials budgets have been especially hard hit. Teachers' salaries have plummeted in purchasing power, aggravating the existing shortage of competent science, mathematics, and technical teachers. Money needed for the updating and rewriting of science and mathematics curricula has not been made available.

Meanwhile, states face financial difficulties as a result of periodic downturns in revenues as well as heavier social spending obligations and "infrastructure" expenses. The combined state and local contribution for local schools dropped from 30 percent of all state-local expenditures in the early 1970s to less than 25 percent in 1981, a trend reflecting declining school enrollments and a reallocation of resources to health, hospitals, and welfare.[10] The federal share of the bill for public elementary and secondary education has dropped to its lowest point in a decade (6.2 percent). Between 1980 and 1985, the federal education budget was reduced by 20 percent in constant dollars.[11] States are in a poor position to provide long-term additional support for public education. Moreover, few states have been able to provide universities and other post-secondary institutions with the equipment, research facilities, and competitive faculty salaries needed to develop or expand technical education programs.

Money is not the only problem. Ironically, just as the sophisticated microelectronic devices heralding a new industrial era were being developed in the United States during the 1960s and 1970s, the academic preparation of American students was sliding downward. Student achievement began to drop in 1963, accelerated in the 1970s, but showed slight signs of improvement for some groups in the 1980s.[12] These facts were widely disseminated with the publication, in May 1983, of *A Nation at Risk* by the National Commission on Excellence in Education. The Commission summarized some of the indicators that characterize the dismal state of American education: 13 percent of all seventeen-year-olds and perhaps as many as 40 percent of minority youth are functionally illiterate; the performance of the average American student in mathematics and science does not compare favorably with that of their counterparts in several other industrialized countries; the scores of both high school and college students on a variety of standardized achievement tests have dropped since the early 1960s yet average grades have risen; and the number and proportion of students scoring 650 or higher on College Board SATs has declined significantly. Moreover, a substantial percentage of students lack "higher order" intellectual skills—for example, being able to solve a mathematics problem that has several steps in it or writing a persuasive essay.[13] Clearly, the solid scholarly preparation that students will need in order to cope with changing economic circumstances is sorely lacking.

There are a host of reasons that help explain why student performance has fallen off since the 1960s. A significant number of students shifted out of the academic and vocational education programs to the amorphous and less rigorous "general" track. The percentage of the high school population enrolled in the "general" curriculum grew from 12 percent in 1964 to 42 percent in 1979. Elective courses in core areas of the high school curriculum (e.g., the substitution of courses such as Science Fiction, Radio/Television/Film, Mystery and Detective Story for the standard literature and grammar courses) proliferated, and there was a drop in the number of academic courses taken.[14] The amount of time high school students spent on homework declined during the 1960s and dropped even further in the 1970s. (Seniors averaged 4.76 hours of homework per week in 1972 compared with 4.21 hours in 1980.)[15] Significant rises in absenteeism in schools during the 1960s and 1970s, increases in automatic promotions to failing students, and a drop in textbook standards have also been cited as causes of the decline.[16]

Other factors implicated in achievement declines include: grade inflation, the introduction of "minimum competency" examinations in many states that has led to a reduction of educational standards to the lowest common denominator, and a reduction in college admissions standards.[17] A decline in teacher quality is also considered a factor, particularly in the areas of secondary mathematics and science where a chronic shortage of qualified faculty has become exacerbated in recent years. The decline in achievement of teacher-trainees and newly prepared teachers has been even sharper than the drop in the academic performance of the general college student population.[18]

Nonschool factors have also been singled out as causes in waning achievement: the significant increase in television watching and decline in study aids at home, such as reference materials (despite the fact that parents' average educational level increased in the 1970s)[19]; the substantial growth in single-parent families and in families where both spouses work; the development of political and cultural shifts, which may have reduced student motivation; and the growth in the number of hours high school students spend in paid employment.[20]

In the wake of the spate of critical evaluations of U.S. schools, a wave of initiatives for upgrading American education in the 1980s are being proposed and implemented. More than 200 statewide study groups and commissions have recommended reforms of various kinds. Many of these address the students' deficiencies in mathematics and science education. Most states have increased high school graduation requirements in core academic subjects and have raised entrance requirements to their public colleges and universities. In many states, textbooks and curriculum materials are being upgraded, teachers' salaries are being raised modestly, and standards for teaching performance are being tightened. In some localities, the amount of homework given students has been increased and the school day and school year, which are short in the United States compared to other industrialized countries, have been lengthened. Thus, there are concrete changes taking place that appear to be moving educational institutions in a direction more congruent with economic trends.

However, although a number of educational reforms are actually being carried out, many alterations remain only in the discussion stage. And as has been proved in the past, education is a vast decentralized enterprise that is relatively resistant to change. Further, most of the changes being recommended will require substantial infusions of public funds. Estimates of what it would cost to implement the changes recommended by the National Commission on Excellence in Education range from $14 billion to $24 billion annually. Despite the fact that several federal intervention efforts have had positive effects on school achievement for specific categories of students, President Ronald Reagan has insisted that federal aid to public education has failed.[21] His administration, which has promoted the view that states rather than the federal government should assume leadership in resolving educational deficiencies, will not move to put into practice the recommendations of its own National Commission.

There has been a flurry of legislative activity in the U.S. Congress to shore up education. In 1984, it authorized (but then did not appropriate) funds for college scholarships for 10,000 students in the top 10 percent of their high school class who planned to train as teachers. These recipients pledge to teach at least two years after college graduation for every year they receive the fellowship. Also in 1984, Congress passed the Education for Economic Security Act, a two-year $1-billion initiative to support mathematics, science, and technology at the secondary school level. But only $100 million was actually appropriated to states and school districts for 1985, an insignificant sum considering the dimensions of the problem. Congress is preoccupied by attempts to reduce the massive federal budget deficit, so continued freezes or real cuts will most likely continue for education programs. It is doubtful that major and sustained initiatives will develop as long as support from the Reagan administration remains lacking.

State governors, believing that educational upgrading is essential for economic development, have moved to fill the vacuum in leadership and funding created by President Reagan's stance on educational issues.[22] State governments have the legal authority to influence a number of recommended changes and are raising educational standards and expectations. The economic recovery of the mid-1980s allowed many states to increase allocations for education, and tax increases passed in others helped pay for school improvements. States have been especially aggressive in setting up and expanding engineering, computer science, and post-secondary technical programs. But many lack the funds to fully implement important changes, such as significantly improved teacher salaries or major equipment purchases that would require extensive budgetary outlays. Overall, the new monies flowing into education are insufficient to pay for long-term reform efforts needed over a five- to ten-year period. Further, states and localities have had to take on new budgetary burdens as the federal government in the Reagan era transfers responsibility for some domestic programs to them. And school costs will rise as student enrollments increase during the remaining years of this decade and into the 1990s.

Moreover, unlike the late 1950s, during the post-Sputnik furor, the constituency for educational expenditures at the elementary and secondary level has eroded, with parents of children in the public schools now forming only 27 percent of the adult population compared with 39 percent in the early 1970s.[23] Declines in American education have now been fully publicized, but whether the political will exists to stem and reverse the decline remains in doubt.

Industry Support for Education

There has been a rapprochement between private industry and educational institutions in the 1980s, following two decades of relative distance from one another. A number of task forces and commissions studying American education have recommended stepped-up cooperation between the private sector and educational institutions as a way of solving some of the problems facing schools and colleges. It is reasonable to assume that institutions of learning and employers of their graduates would work in a variety of collaborative ways. This is especially true now when so many high school graduates are deficient in such fundamental competencies as writing and computation. Employers have to spend increasing amounts of time and money to provide remedial training to their workers because deficits in basic skills reduce productivity on the job. A recent report of a survey of business and education leaders by the Center for Public Resources concluded that the time is ripe for enhanced industry–public education cooperation:

These . . . survey findings evidence the importance of the basic skills problem, and the willingness of business, union, and school system leaders to implement more productive and cooperative methods of problem resolution. The incentives to business are clear-cut and relate directly to market performance. The public benefits are also clear. The linkages among economic growth, work force employability, and worker productivity are tightly forged. Basic skills competency is a common factor in the linkage. . . . The prospects, therefore, for mobilizing long-term, sustained public-private partnerships to assist public education are excellent.[24]

The mutual need of both institutions of higher education and industry, especially high technology industry, for tighter links seems equally obvious. Companies claim there is a shortage of highly trained technical professionals at a time when colleges and universities cannot afford increasingly sophisticated training equipment and expanded staff to handle burgeoning student interest. Companies welcome access to the expertise of university researchers. The financial needs of academe and the work force and research imperatives of companies would seem to encourage close and enduring partnerships of various kinds.

There has been, in fact, an undeniable upswing in industry-education cooperation in the late 1970s and 1980s. Certain types of collaboration, such as corporate representation on employment training councils and state vocational education councils, have been institutionalized by federal law. In addition, companies are providing a variety of resources for schools and colleges. There are now a considerable number of corporate–public school programs especially in the nation's largest cities. Many of these efforts revolve around training hard-to-employ youth. Another common focus is career awareness education, often coupled with plant tours. Business people have historically been involved in the planning and evaluation of vocational education curricula. Computer education is now providing a new point of common effort and interest. A number of firms are donating computers to schools and are providing other sorts of support to spur computer literacy. Again, the mutual needs are obvious. Schools have difficulty affording the huge outlays required for adequate hardware and software. Companies, for their part, want access to school and home markets as well as the expansion of the pool of computer-trained citizens who will be available as potential employees and consumers.

At the college and university level, links with companies are even more pronounced. For example, cooperative research efforts have grown significantly in the last few years. Industrial parks on college campuses are being established, collaborative microelectronics centers are being built, training programs for company employees involving campus resources are growing, and corporate donations to academe are rising. The bulk of cooperative activity between universities and all kinds of business firms is occurring in the fields of engineering and biotechnology.[25]

However, for all the talk of "the new partnership" between educational institutions and industry, relations between the two are uneven and tentative. The much-heralded alliance is occurring primarily at the upper end of the educational system—at the elite colleges and universities. Significant barriers to cooperation still separate most schools from the corporate world. Relations between public schools and business are usually fragmentary, weak, and of short duration. Executives are far more willing to donate funds, personnel, and equipment to higher education because funds can be easily targeted to specific programs. Business people want a quick return on their investment, something public schools can rarely deliver. Communication between the companies and nonelite colleges and universities is looser and less satisfying than that which exists between corporations and the more prestigious academic institutions. Most of the publicity and activity in the area of company-university relations centers on a relatively small number of schools in America's vast higher education establishment.

Corporate support alone cannot compensate for reductions in government funds for education or provide for the large increases in funds needed to bring American technical education up to a state-of-the-art level. Indeed, at the public school level, schools receive ten times as much money from cake sales and ticket sales as they do from corporate contributions.[26] Business groups acknowledge that their role can only be a supportive, supplementary one. Despite a general "small government" philosophy, some industry associations have gradually reached the conclusion that significant public funds are needed. A small but increasing number of prominent executives are serving on educational study groups and task forces which have decided that educational reforms must be accompanied by increased public fiscal support. Some corporate leaders and groups are joining or even leading coalitions seeking significant changes to upgrade American education. But this fledgling alliance is fragile. In some states, corporate commissions and study groups have promoted educational reform in theory but then have withheld support from the reform legislation when it necessitated a tax increase.

There are some grounds for being wary about the renewal of interest and influence in education by the private sector. The sharing of information among academics at some major universities is being affected by the restrictions inherent in corporate-sponsored research. Research priorities in some scientific fields are becoming skewed by commercial considerations, and the upgrading of engineering and other technical programs needed to meet industry work force requirements is being done at the expense of liberal arts curricula on some campuses. Training programs in community colleges and technical schools are often too narrowly focused in order to serve the needs of specific employers. And while business participation is certainly desirable in selected areas of education, it is unsettling to hear business representatives argue, as one did at a press conference following the passage of the 1984 Vocational Education Act, that "this law puts control over vocational education where it belongs—in the hands of people with jobs."[27]

But despite the dangers that accompany some collaborative efforts, the greater peril lies in continuing corporate indifference to the needs of American education. Support for education is simply not a high priority item for most firms and business associations. Business executives want a strong, thriving educational system just as they would like to see other efficient, attractive public services. Thus far, however, they, are not willing to throw their considerable weight behind essential public expenditures for educational and training programs. Yet the building of a revitalized and technologically oriented economy cannot be constructed on a crumbling educational foundation. The "second industrial revolution" must be accompanied by an educational renaissance as well.

2 The Role of Our Schools in a High Tech World

The advent of a "new industrial revolution" has sparked a reappraisal of the state of America's educational system. Educational upgrading is now high on the action agenda of many political leaders, especially governors, who look on education as a tool of development in a changing economy. They have been joined by business leaders and educators in pressing for a broad academically stringent education at the pre-college level and a greater commitment of resources to technical education in post-secondary institutions. Even though the introduction of new technologies into the workplace may actually lower skill levels required in selected jobs, these political and business leaders and educators are correct in their belief that well-educated workers are more likely to adapt to and survive economic shifts.

Clearly, an economic shift is upon us now. The first industrial revolution saw machines replacing human muscle power; in the "microelectronics revolution" computers are being substituted for human brainpower. As a result, modern industrial societies are now "information-" or "knowledge-based" systems since they are so dependent on the efficient flow and use of information, and since the computer can often outperform the human brain in the processing of information. Computers are able to handle an enormous volume of information in a systematic and inexpensive way and their miniaturization has made it possible for them to be integrated into a wide variety of consumer and industrial products.

In the first half of the 1980s, the average American citizen became aware of the pervasiveness of the changes wrought by modern electronics. Americans now wear electronic digital watches, use hand-held calculators, conduct banking with automatic tellers, watch their food and books checked out by laser terminals in grocery stores and libraries, and sit in front of video display terminals at work. Computers are rapidly penetrating the home and the school, and discussions of computer products are now a part of people's daily conversation.

The development of modern electronics and the subsequent computer age have come about as a result of advances in semiconductor devices, which transmit, redirect, amplify, and store electrical impulses. The invention of the semiconductor integrated circuit by Robert Noyce at Fairchild Semiconductor and Jack Kilby at Texas Instruments in 1959 ushered in the era of microelectronics. These integrated circuits consisted of a network of transistors, capacitors, diodes, and resistors linked by thin aluminum strands that were placed on a chip of silicon smaller than a fingernail.

The development of the microprocessor by M. E. Hoff (the placing of the entire central processing unit of a computer on a single chip) at Noyce's Intel Corporation and the microcomputer in 1971 (a central processing unit chip and storage chips mounted on a board) by that same firm further advanced this revolutionary technology. The number of components on a chip expanded rapidly while prices for the chips dropped precipitously. Between 1962 and the early 1980s, semiconductor devices dropped in price about 35 percent a year.[1] More astonishing price-cutting has occurred since then. For example, large-scale (256K) memory chips dropped in price from as much as $45 apiece to less than $5 during 1984 and 1985. As one writer put it, "The microcomputer . . . is one of those rare innovations that at the same time reduce the cost of manufacturing and enhance the capabilities and value of the product."[2]

There is a tremendous amount of awe and excitement about the capacities of the computer on a chip and the unprecedented rate at which its scale of integration has proceeded. The fact that as many as a million transistors can fit on a chip—up from ten in 1960—is difficult for the human imagination to grasp. Even the late Frederick

Terman, former dean of Stanford's School of Engineering and the "father of Silicon Valley," was once quoted as saying, "I understand how these things work but I still don't believe it . . . I simply disbelieve that anyone can put 5,000 transistors on a little silicon wafer one half-inch square."[3] Chips themselves are becoming more complex and the range of tasks they can perform is expanding. The applications of microprocessor technology in home and office products has generated enormous popular interest and enthusiasm. Computer clubs, magazines, and conferences for microcomputer users abound. American consumers, including children, are fascinated by electronic gadgets of all kinds. High technology firms for the most part have generally been viewed in a favorable light by the press and government leaders.

Only recently, however, has it become clear how profound the ramifications of microelectronic technology will be. Numerous jobs are being changed in fundamental ways as microprocessors invade the workplace. The impact is being felt first in white-collar offices as large segments of secretarial and clerical work become automated by word processing equipment. The printing industry is being transformed by electronic typesetting and photocomposition and will be affected by electronic mail and newspapers. Banking is becoming increasingly automated. Retailing is being changed by "smart" terminals while some aspects of education, notably industrial training, are utilizing computers as teaching tools. Robots are just beginning to make their presence felt in the U.S. in traditional industries, such as automobile manufacturing, and will be utilized more extensively in manufacturing in the 1990s. The replacement of the traditional watch and clock by electronic devices has brought about vast changes in that industry. Microprocessors are increasingly being used in textile production. Architects are using computer-aided design processes; lawyers have access to data banks; and doctors can use electronically sophisticated diagnostic techniques. These are only a few of the many ways in which microelectronics is transforming industrial, professional, and service occupations.

What Is High Technology?

While high technology is the buzzword of the decade, there is no universally agreed-upon definition of the term. Some analysts argue that the pervasiveness of innovation is the hallmark of a high technology firm. Lynn E. Browne, an economist at the Federal Reserve Bank of Boston, puts it this way:

High technology is not a particular industry or product. . . . Rather, high technology industries are industries which are characterized by change and innovation; they are industries in which new products and processes are continually being generated. If an industry ceases to develop technologically, then it should not be considered high technology.[4]

The executive director of Connecticut's High Technology Council, James P. Fenton, whimsically defines high technology as "any industry that is going to create jobs in the 1980s and 1990s."[5] According to the Massachusetts Division of Employment Security and Department of Manpower Development, high technology industries have the following characteristics: a highly skilled employee base, including a high concentration of scientists and engineers; rapid growth rates; high ratios of research and development expenditures to sales; high value-added products; and worldwide markets for products.[6] These labor-intensive industries include the manufacture of drugs; ordnance and accessories; office computing and accounting machines; electrical and electronic machinery, equipment, and supplies; guided missiles, space vehicles, and their parts; miscellaneous transportation equipment; and measuring, analyzing, and controlling instruments, including photographic, medical, and optical goods. Certain service industries are also considered to fall into the high technology category—computer programming services, commercial research and development laboratories, some business management and consulting services, engineering and architecture services and some nonprofit educational, scientific, and research organizations. The Massachusetts High Technology Council (MHTC), the major trade association for the industry in the state, defines its membership in similar terms.

The U.S. Bureau of Labor Statistics has found it helpful to create three definitions of high technology employment. They have defined one set of industries as being high technology if the proportion of technology-oriented workers (engineers, life and physical scientists, mathematical specialists, engineering and science technicians, and computer scientists) is at least one and a half times the average for all industries. This is the broadest definition of high technology and includes large portions of the chemical industry, petroleum refining companies, the motor vehicle and equipment industry as well as machine-producing companies. This group employed 13.4 percent of American workers in 1982. A second and much more restricted definition of high technology includes only those firms whose ratio of research and development expenditures to net sales was at least twice the average of all industries, a group that included only 2.8 percent of the U.S. work force in 1982. A third definition, which included companies employing 6.2 percent of the labor force, combined the utilization of technology-oriented workers (whose percentage had to be equal to or greater than the average for all manufacturing industries) and the ratio of research and development expenditures to sales (close to or above the average for all industries), and also added computer and data processing services and research and development laboratories.[7] As new technologies pervade traditional industrial work sites, and as the rate of growth and innovation decreases in some electronics and computing firms, it will become increasingly difficult to spell out a precise definition of high technology business.

High Technology, Education, and Economic Development

There has been a tremendous upsurge of interest in the 1980s on the part of governors and other political officials in attracting high technology companies to their states and regions. Much of this concern stems from the fact that employment in such traditional key manufacturing sectors as automobiles and steel has declined dramatically in the 1980s. The fastest growing part of goods-producing industry in recent years has been the high technology sector. High technology industries in 1980 accounted for 19 percent of total manufacturing employment and, according to one federal estimate, as much as 75 percent of the growth in manufacturing employment from 1955 to 1980.[8] A 1984 study for the National Science Foundation found that high technology firms could take credit for 42 percent of the net growth in manufacturing jobs between 1976 and 1980.[9] Moreover, the rate of job growth in high technology industry during that period was considerably higher than that of the traditional "smokestack" manufacturing sector.[10] In a different analysis, however, the Bureau of Labor Statistics shows that, depending on the definition of high technology that is used, the high technology industries accounted for only 4.7 to 15.3 percent of new jobs in all employment sectors (not just manufacturing) between 1972 and 1982.[11]

But this overall figure masks greater gains in several states, such as Massachusetts where high technology employment, according to the Bureau of Labor Statistics study, directly generated from 18 to 35 percent of all new jobs in the state between 1975 and 1982. Economist Marshall Goldman argues that since the creation of these jobs had the effect of expanding employment in services and other regular industries as well, almost half of all job growth in the state between 1972 and 1982 could be explained by high technology firms.[12] Clearly, a few states and metropolitan areas have benefited greatly from the presence of high technology complexes. High technology businesses generally have higher growth rates and plan to invest more heavily in plant and equipment than other manufacturing firms.

With these data in mind, it is not surprising that regions, states, and some metropolitan areas are engaged in a fierce contest to retain and lure high technology firms. ". . . Every place in the country has suddenly gotten the idea of high technology," claimed Thomas A. Vanderslice, the president of GTE Corporation.[13] At least thirty-three states are in the process of spending a total of $250 million to develop jobs in computers and electronics.[14] Bidding wars have broken out among states and cities in their efforts to attract industry. A host of inducements are being offered to firms: tax breaks, public venture capital funds, research parks, and "incubator" facilities for start-up firms, among others.

Officials in Austin, Texas, went beyond even these inducements in their successful bid to attract the Microelectronics and Computer Technology Corporation (MCC). MCC is a research collaborative of twelve leading firms whose aim is to outstrip government-funded research and development efforts in Europe and Japan. Its offer included rent-free use of both temporary research facilities and permanent offices and laboratories in a twenty-acre research park, $20 million in single-family home mortgage loans at reduced interest rates, up to $500,000 in relocation expenses of new employees, and the use of a Lear jet for the MCC director and his staff for two years as they recruited personnel.[15] Economic development officials from various states send scouts and delegations to try to woo high technology firms away from their current centers. This activity is spurred by the fear of continued high levels of joblessness throughout the 1980s, regardless of whether the economy is strong or not.

Governors are probably more sensitive to high unemployment levels than are federal political leaders since the falloff in revenues and increased demand for social services accompanying unemployment invariably wreak havoc with state budgets. State and local governments have formed numerous task forces and commissions to develop plans for economic growth. Indeed, concern with economic development has been more pronounced at the state and regional levels than it has been at the federal level, at least during the tenure of the Reagan administration.

The economic development plans being written at all levels of government stress the importance of a well-educated labor force. This theme has been paramount in the inaugural addresses of a number of governors in the 1980s. Consider, for example, an excerpt from Governor Michael Dukakis's 1983 address in Massachusetts:

Three great forces that changed the world were forged here in Massachusetts: the American Revolution, the Industrial Revolution, and the High-Tech Revolution. Once a barren wilderness and a small area with few natural resources, Massachusetts has been a powerful force for change and progress because this state has nurtured and harnessed the most powerful resource we have: our people. Our wealth has been in the quality of our life; in a frugal and industrious and intellectually curious and educated citizenry; and in the power of knowledge and discovery that is our only renewable and inexhaustible resource.

Policymakers repeatedly argue that businesses need adequately trained and educated workers at all levels of employment. This argument is stressed most forcefully in the plans of those areas trying to retain or attract high technology firms, particularly those stressing research and development. A 1982 study by the Joint Economic Committee of the U.S. Congress of factors that influence the regional location choices of high technology companies found that the availability and skills of a labor force ranked first among firms' location considerations, followed closely by labor costs. Proximity to academic institutions ranked fourth among the twelve factors considered.[16] Unlike more traditional manufacturing concerns, which are concerned with nearness to markets, natural resources, and transportation, high tech firms are regarded as relatively footloose

"industrial vagabonds." Instead, the "knowledge-intensive industries" are more concerned with locating in areas where concentrations of technical brainpower exist. Such firms prefer locating in
close proximity to a top-flight university whose engineering and
computer science graduates they can hire and whose graduate programs are available for their employees. ("Because we're a high tech
industry, education is very important," claims a spokesperson for
Motorola. "We have to be within 30 minutes driving time of an
evening graduate school.")[17] Indeed, in the case of Stanford University and the Massachusetts Institute of Technology, the Silicon Valley and Route 128 electronics complexes never would have
developed in the first place. High technology companies also like access to a good community college system that is sensitive to their
training needs.

Although good public school systems are not usually of direct importance to the firms' labor supply, they figure in the formulation of
an elusive and informal "quality of life" index that is crucial in plant
location decisions. For example, according to a Honeywell vice president, Ed Roach, high taxes and high labor and facilities' costs in
Minnesota are offset by the state's "reputation for strong public
education systems and cultural amenities which make it relatively
easy for the firm to attract engineering and technical talent to its
Twin Cities operations."[18] Impressionistic evidence gleaned from
corporate interviews for this study indicates that such factors are
quite important. In a recent analysis of Chicago's bid for high technology firms, one economist claimed that "a critical factor relating
to the Chicago area's ability to attract and keep jobs and firms is the
quality of public education, which is lousy."[19] Proximity to good
public schools becomes an important factor in selecting a specific site
for a plant once a firm has settled on a general region for expansion
or development. In the 1982 study by the Congressional Joint
Economic Committee, high technology firms ranked the school factor as seventh in importance (out of fourteen) with 71 percent of those
responding claiming that good schools were a significant or very
significant consideration in selecting a site within a region.

Industry trade associations, particularly the American Electronics Association, have stressed the need for skilled technical
labor. They have publicized the problems plaguing engineering education in the United States and have encouraged public and private
higher education to expand engineering and computer science pro-

grams. The associations have also highlighted the need for more and better technician training efforts. Some R and D branch plants of high technology companies have located overseas (in Israel and Scotland, for example) in part to be near elite technical education research and training facilities. A number of industry executives have warned that the future of high technology in America is dependent upon an increased supply of personnel with strong training in mathematics, science, and technical subjects. Even if other nonindustry analyses of labor force needs sometimes differ from those of business leaders, it is not surprising that government officials, desperate to attract companies, feel obliged to adhere to industry predictions.

Therefore, as a part of economic development packages, governors, mayors, and state legislatures are scrambling to reorient their educational systems in a more technical direction. Between 1981 and 1984, more than half the nation's governors appointed committees to provide direction in developing educational and other programs that would promote high technology industry. Some states have developed new schools of engineering or have expanded old ones. In the hotly contested effort among fifty-seven cities to attract MCC, Austin, Texas, won in part because it promised more than $20 million for teaching positions, research support, laboratory equipment and graduate fellowships for computer science and electrical engineering programs at the University of Texas.[20]

Interactive telecommunications systems are being subsidized by some states to provide on-site graduate programs in engineering and computer science for engineers in industry. Other states have developed customized technician and assembler training programs for companies through community colleges or vocational education centers, or have instituted and exanded two-year degree programs in technical fields. Industrial parks located on university campuses are springing up. Economic development is increasingly invoked as a reason for upgrading public schools as well, a push that is leading more and more states to stiffen high school graduation requirements, teacher certification standards, and admissions criteria to public colleges and universities.

The incorporation of educational changes into economic development schemes is not limited to the United States, a fact not lost on American political leaders. Scotland, for example, is promoting its universities and fifty-five technical colleges as part of a support system for high technology industry. A microelectronics institute has been established on the campus of the University of Edinburgh, and the Scottish Development Agency estimates that 40 percent of Scotland's scientific and technical researchers are studying very large scale integrated circuit technology (VLSI), optoelectronics, and artificial intelligence.[21] A number of British universities are moving to establish research parks where faculty will collaborate with high technology companies.[22]

Ireland has made a major effort to expand engineering, electronics, and computer science courses in its established universities, and the government has set up nine technical colleges to increase the flow of technically trained personnel needed by newly arriving foreign electronics and computer firms. The government has also created training centers for less-skilled technicians and two new National Institutes of Higher Education to educate workers for these firms. Singapore has taken steps to train more electronics engineers and computer software specialists, hoping to capture some of the research and development of computer software from foreign firms and to establish more indigenous firms serving that region of the world. The government has moved quickly to train greatly increased numbers of technical professionals and has introduced computer training into all levels of public education. Thus, both domestic and foreign governmental units are moving to refashion their educational systems in ways that will be more congruent with the needs of high technology firms.

A High Technology Future?

When this research was first begun in 1980, there was an unquestioned assumption among leaders in both industry and education in the United States that there would be a vast expansion of job openings throughout the 1980s for highly skilled workers in high technology companies and related service industries. It thus made a good deal of sense to think about and advocate methods to improve the quality of technically oriented educational programs and to increase significantly the quantity of newly trained engineers and other technical personnel.

Since that time, several factors have challenged this point of view. First, high technology companies were hurt by the general economic recession of the early 1980s more than they anticipated, and a number of the firms in dynamic, high growth sectors had to lay off workers or place a freeze on hiring. Periodic slumps in the semiconductor industry, plagued by price-cutting and competition as well as a drop in orders, have been particularly severe. Employment in electronics rose dramatically in late 1983 and 1984 but decreased substantially in semiconductor firms and computer manufacturing in 1985. This drop was especially worrisome since the rest of the economy was no longer in recession.

Second, it became apparent that the new technologies increasingly being introduced into a spectrum of workplaces were often either displacing workers or preventing further job creation. As both high technology and traditional manufacturing firms began to recover from the recession, they did not rehire all their former workers, in part because of new labor-saving technologies. Third, although the electronics and computer industries are still projecting tremendous growth as a proportion of their current size, the absolute numbers of new jobs to be generated by the 1990s does not appear to be nearly enough to employ even those manufacturing workers who have recently lost their jobs. Fourth, as competition among various regions of the U.S. for a piece of the high technology pie intensifies, more and more business analysts and government leaders realize that these firms will be the economic salvation of only some of the nation's cities and regions. Fifth, there is evidence that a variety of computer-driven technologies will downgrade the skill levels of some jobs rather than upgrade them.

With these sobering trends in mind, some economists, including Henry Levin and Russell Rumberger at Stanford University, are arguing that the advent of high technology does not require a redirection of American education along more technical lines for the work force as a whole.[23] Their arguments fit in with a more general critique of the work force requirements of high technology industry that emerged in 1983 and thereafter which questioned the ability of the industry to provide plentiful and meaningful work. This perspective is at odds with that of numerous industry and government officials and commissions, who link economic development to the growth of high technology firms, and who further connect such growth to the upgrading of technical education, including the teaching of mathematics and science.

The debate unfolding about the nature of future work force

needs will be a central political issue for the remainder of the twentieth century. Several major questions must be addressed: Will the new technologies create more jobs than they destroy? What kinds of jobs will be eliminated and what will be the skill requirements of the new jobs? Will the introduction of computer-based products reduce or increase skill levels of currently employed workers? Those who make decisions about the future of education and training must have some familiarity with the issues and evidence bearing on these questions.

High technology companies today employ about 3 to 13 percent of the nonagricultural labor force in the United States, depending on which definition of high technology is used. A 1983 U.S. Bureau of Labor Statistics study argues that somewhere between 1 and 4.6 million new jobs in high technology industries will be created between 1982 and 1995, accounting for 3 to 17 percent of total job growth during that period.[24] A 1983 survey conducted by Data Resources, Inc., for *Business Week*, using a somewhat different definition of high technology, shows that only 730,000 to 1 million jobs will be created in the industry by 1993, less than half of the 2 million jobs lost in manufacturing since 1980.[25] (As one AFL-CIO official put it: "High tech doesn't have an answer. All those people can't leap from that abandoned industrial base onto this little tiny silicon chip for our future. There's not enough room.")[26] Even though the industry will create more jobs than other manufacturing sectors and will grow at a faster rate relative to other industries, its small current base and its own propensity to improve productivity and automate production means that it cannot be counted on to generate vast numbers of jobs.

The term "jobless growth" has been coined to describe the phenomenon of a rising dollar output of firms without a corresponding rise in employment.[27] This expression is particularly apt when examining work force trends in high technology firms. *Business Week*, citing the Data Resources study, puts it this way:

. . . Output per high tech worker in 1972 dollars is expected to rise by 46 percent between 1983 and 1993, nearly double the projected increase of 24 percent for manufacturing and 23 percent for services in the same period. This means that while dollar output in high tech industries will grow 87 percent over the next decade—from 7 percent of gross national product to 10 percent, or $206 billion—the number of workers needed to produce this increase will need to rise by only 29 percent.[28]

These are not merely predictions about future trends. Productivity increases, fueled by plant automation, are already resulting in a lower demand for assembly-level workers in semiconductor plants and other segments of the electronics and computer industries. At Shugart Corporation in Sunnyvale, California, for example, three workers today turn out 1,000 printed circuit boards per shift, a task that used to employ twenty workers per shift before the assembly line was automated. ("We've made labor a nonissue," commented one of the company's managers.) A 1984 in-depth investigation by *San Jose Mercury News* reporter Pete Carey found that the automation of Silicon Valley electronics companies is proceeding so rapidly that most semiskilled and unskilled jobs in the industry will disappear by the mid-1990s.[29] Certain types of work performed by electronics test technicians and maintenance technicians are also becoming more automated in some firms, reducing the overall demand in some sectors of that occupation. These trends will continue as computers become better able to perform a variety of mechanical and mental tasks.

Moreover, the small size, versatility, low price, and tremendous power of microelectronic devices means they can be applied flexibly and successfully in almost all sectors of manufacturing and service work. The job-displacing effects of the new technology have already become obvious and the "automation debate" has become a central political issue in Western European countries, the United Kingdom, and other industrial democracies. Foreigners are surprised that the debate over this issue is only beginning in the United States, although the initial controversy over technology and unemployment in the early 1950s took place in this country. The fact that the automation scares of that period did not come true does not mean that new predictions about unemployment should be dismissed. The small computer chip of today can be integrated into the workplace far more easily than the large clumsy and relatively labor-intensive computers of the early 1950s, and its introduction comes at a time when economic growth has been stagnating.[30] Further, as automation expert Harley Shaiken of MIT notes, automation in the past usually occurred in only a few economic sectors at a time, as was the case when agriculture was affected; but we have a situation today where almost all workplaces are being affected by technological displacement simultaneously.[31] The implications for structural unemployment rates are ominous.

A number of studies have demonstrated that as a result of the introduction of new technologies, a substantial loss of jobs has already occurred and job creation potential is being reduced. Both U.S. and international reseach efforts have noted significant losses in office work and in industries where microelectronic components have replaced mechanical and electromechanical parts, such as the watchmaking, cash register, office equipment, and telecommunications equipment industries. Similar reports of labor displacement come from a wide range of other occupations, including printing and typesetting, insurance and banking, textiles, automobile manufacturing, telecommunications service, machine tool trades, and wholesale and retail services.[32]

Other studies, however, argue that new technologies are being applied in work settings so slowly and unevenly that their overall impact on employment levels during the 1980s and 1990s will not be pronounced. This is especially true in the United States, where economy-wide employment is growing, in contrast to the lack of job creation in Western Europe. Analysts do point out, though, that selected industries and regions (e.g., Ohio, Michigan, and Illinois, which are heavily involved in the metalworking industry) may suffer substantial job losses as computerized manufacturing automation is adopted.[33] And as microelectronically powered devices become more flexible and sophisticated during the 1990s, their labor-displacing effects may become dramatic early in the next century.

But as computer and electronic technologies destroy jobs, they are creating others at the same time. Not only have whole new occupational categories been created within high technology companies themselves, a host of related service industries have sprung up. Moreover, new jobs are created within traditional manufacturing and service occupations as they adopt technical innovations. And high technology companies by their very presence have a multiplier effect on surrounding businesses. The question is not whether the new technologies will create jobs, but whether or not there will be sufficient job creation to offset jobs displaced by the same technologies. The evidence on this point is certainly not all in, and we are still in the early stages of computer and electronics product development and applications. There are many research problems associated with attempting to forecast work force trends, and U.S. data-gathering agencies are hampered by reductions in their staffs as a result of budget cuts. No one really knows what the ultimate employment effects will be.

It is important to point out here that the degree of unemployment due to labor-saving devices will be determined by a variety of political and economic factors, including employees' control over the introduction of new technologies, the length of the work week and work day, general economic conditions, and governments' political commitment to full employment. Obviously, for example, if work weeks are shortened as labor savings occur, job displacement effects will be reduced. Countries with strong trade union movements are most likely to have contracts that reduce workers' vulnerability to automation. Also, it is essential to bear in mind that even though certain occupations may shrink in size, there is still plenty of socially necessary labor-intensive work to be done in American society—the building of homes and public transit systems; the rebuilding of the nation's roads, dams, sewers, and bridges; the provision of social services; and the installations of energy-saving products, to name just a few. The question is whether sufficient public funding will be made available to accomplish these tasks. That is a political issue. In any event, it is erroneous to come away from the automation debate with the notion that "there won't be any work for people to do" in the post-industrial age.

Aside from the issue of whether there will be more or fewer jobs as microelectronic devices permeate the workplace, there is the question of what skill levels will be required for the jobs that remain and those that are created. One school of thought promotes the view that existing jobs will be "deskilled," becoming more controlled, monotonous, and simple-minded. Proponents of this view argue that skill levels of many current jobs will be downgraded. They cite clerical work, retail jobs, printing, machine tooling, and computer programming, for example, as computers take over more complex mental tasks. Writers, Harry Braverman for example, look at the history of production in industrial societies and claim that work has consistently been broken down into smaller parts.[34] This finer division of labor robs employees of autonomy and satisfaction and shifts control increasingly into the hands of the employer. According to this school of thought, the adoption of computer-driven devices will accelerate this process for most workers since it will simplify and segment their work even further. Recent case studies of several occupations, including office workers and machine operators, confirm this line of thinking.[35] Levin and Rumberger also dispute the view that the pervasive introduction of computers will require legions of highly skilled workers to run them. They use the analogy of the automobile to illustrate that a person does not need to understand how something functions (how the car works) in order to operate (drive) it, just as most computer users will not need to know what makes computers work.[36]

Those on the more pessimistic side of the deskilling debate also argue that the areas of high job growth in the economy will be those jobs requiring little discretion and training. Levin and Rumberger and others cite forecasts from the U.S. Bureau of Labor Statistics showing that the major absolute growth areas of employment will be in low-skilled work: janitors, nurses' aides, sales clerks, cashiers, and waitresses. They point out that "these five jobs alone will account for 13 percent of the total employment growth between 1978 and 1990. Only 3 or 4 of the 'top 20' occupations in terms of total contribution to job growth require education beyond the secondary level, and only two require a college degree (teaching and nursing)."[37] They cite the projection from the Bureau of Labor Statistics study that while 150,000 new jobs for computer programmers will be created between 1978 and 1990, 800,000 new jobs will emerge for fast-food workers and kitchen helpers during that same time period. Even though the U.S. Bureau of Labor Statistics projects that three of the five fastest growing occupations between 1980 and 1990 will be in high technology fields (data processing mechanics, computer operators, and computer systems analysts), Levin and Rumberger point to the fact that the actual number of jobs in those fields will still be relatively small.[38]

More optimistic analysts suggest that the issue is not so simple. Sociologist Paul Attewell, in his study of insurance company workers, found increased levels of satisfaction and skill among clerical workers now working with video display terminals rather than those working in manual filing tasks.[39] Paul Adler's research on banking employees in France found that their work skill requirements became more challenging and complex. Adler also makes the more general observation that computer users in many job settings must utilize relatively sophisticated judgment and skill:

Computer systems . . . are not entirely standardized—far from it. If small-scale users usually buy packaged software, large-scale users are constantly adapting their applications. The result is that operators have to learn a surprisingly idiosyncratic, constantly evolving company-specific system. . . . They are obviously all the better equipped to do so when they have a basic computer literacy.[40]

The more optimistic observers point to the decline of physically demanding and unpleasant work as machines replace human labor. Looking at work over a long historical period, they argue that it is the most degrading and arduous forms of labor, particularly forms of agricultural work, that have disappeared while newer forms of labor activity have been physically easier and mentally more demanding. A recent two-year British study of the iron castings, metalworking, machine tools, and electronics components industries cautions that the impact of labor-saving devices on skilled manual workers is unclear. While the demand for unskilled workers will certainly fall, they report, the effects on skilled manual workers is less predictable. Although declines in skilled employment will happen in some industries (e.g., iron castings), "new technology will increase the need for flexibility in the use of manpower. This could involve a blurring of current demarcation lines between technicians and skilled workers. . . ."[41] Studies conducted by the U.S. Office of Technology Assessment on the effects of automation in manufacturing also found conflicting trends:

In some cases, the creation of new tasks and the elimination of old ones clearly raises or lowers skill requirements. Often, however, the effect on skill demand is ambiguous, because the skills associated with individual jobs and the average skill level of a company's jobs depend on how well employers understand what skills they really need. By alternating the balance of work between people and machines, programmable automation makes it possible for managers to reallocate work in ways that either raise or lower the skill requirements of jobs.[42]

Like the automation debate, the deskilling contorversy remains unresolved. Few concrete research projects in specific job settings have actually been conducted. Overall, the evidence on the impact of technology in groups of skilled workers is mixed, with skill levels expected to fall on some jobs and rise in others. But virtually all studies agree that jobs for unskilled workers in manufacturing will decline. This means that overall skill levels of workers in manufacturing will increase. While certain low-skill occupations in the service sector are growing rapidly, some of these, such as fast-food work, are often only temporary jobs for youth. Unlike the automation debate, where the weight of the evidence points to net job losses, the evidence thus far on the net effects of deskilling is less conclusive.

The Emerging Consensus

Educational policymakers find themselves, then, in a very perplexing position. Forecasts about the structure and skill levels of future employment are contradictory and unclear, yet they must proceed with fashioning an education and training system for an uncertain future. At a time when industry and many government officials are calling for a vast upgrading of the skills of future American workers, some academic analysts and labor union leaders are arguing that educational levels needed for employment in high growth fields are diminishing. Nevertheless, a tentative and sensible consensus among a variety of researchers and policymakers exists on several key questions that should influence educational directions.

First of all, there is growing agreement that students today and in the future will need a broad, demanding academic education through grade twelve. Business leaders, commonly thought of as promoting only narrow vocational training, are now calling for a broad-based education for all students. A 1982 survey of businesses in the United States found that employers wanted entry-level employees out of high school to have solid training not only in reading, writing, and other communications skills, but in mathematics, science, and problem solving as well. These employers, from a wide range of companies, claimed that entry-level employees, even in such occupations as low-level clerical work, need increased mathematical and scientific competence.[43] A National Academy of Sciences panel composed largely of business executives also has urged that students intending to work directly after high school have a mastery of core academic subjects that will give them the analytic and problem-solving abilities needed to continue learning throughout their work lives.[44]

Other studies of corporate views of education by Teachers College President Michael Timpane and reseachers at The Conference Board found similar results.[45] Urban employers in the Timpane survey complained that few graduates of city schools were qualified for entry-level jobs, that as skill levels for these jobs were rising, educational quality was deteriorating. What they wanted to see was an educational system that taught "basic skills plus."

The basic skills are identified as general skills of reading, writing and computing sufficient to master the technical aspects of entry level jobs, to learn successively more complex tasks over time, and to communicate effectively and exercise good judgment on the job. This is not a narrow "3-R's" conception of the basic educational skills. Nor is it a call for schools to concentrate on providing specific job skills; most corporations believe that they can and should best do this themselves. The call is rather for a set of general though basic skills that are often summed up under such rubrics as "functional literacy" or "employability."[46]

An extensive 1984 report on the current and future impact of computerized manufacturing automation in the United States by the Congressional Office of Technology Assessment stresses the necessity of a strong foundation of academic skills among manufacturing workers, especially in the areas of reading, science, and mathematics.[47]

A 1980 West German study on employment trends also concluded that "more general skills such as abstract thinking, planning skills and ability to work in a team are gaining . . . in significance. This is why general education should put more emphasis on such general skills. . . ."[48] The National Commission on Excellence in Education, drawing on the expertise of a wide spectrum of academic and business opinion, recommended that all students seeking a high school diploma complete the "Five New Basics" by taking four years of English, three years of mathematics, three years of science, three years of social studies, and half a year of computer science.

Although it may ultimately prove to be true that more jobs drop than gain in skill levels as microelectronic technology transforms the workplace, it is manifestly clear that the individual worker is much better off with a sound academic education along the lines suggested by the Commission, rather than the less stringent curriculum in vogue in the last decade. Lesser educated workers will be the most vulnerable to automation of their jobs and the shift of production plants overseas to lower wage areas. Workers may have to retrain several times during their occupational lives, and the better their academic background, the more easily they can shift to different jobs and learn new skills. The experience of West Germany, where students receive a demanding elementary and secondary education, shows that workers can retrain for new jobs more easily when they have good initial education.

A second point of consensus is that educational institutions of all kinds will have to assume some responsibility for retraining veteran workers. As the National Commission on Excellence in Education pointed out, those in the labor force today will constitute about 75 percent of the work force in the year 2000.[49] Their work lives will span a period of incredible economic change. Vocational training centers, community colleges, and four-year colleges and universities will be called upon to develop and expand all kinds of retraining programs. Some of these programs will have to train employees in the newer technologies (although thus far few laid off from "smokestack industries" have flocked to this lower paid and sometimes technically more difficult work). A substantial increase in funds will be needed for these schools if they are to retrain the millions of workers who will lose their jobs as automation proceeds. Funds are also needed in order to bring technical education at the post-secondary two-year schools and the colleges and universities up to a state-of-the-art level. The American higher educational system is not providing adequate training on advanced equipment with enough qualified instructors to serve the needs of engineering, computer science, and other technical trainees currently in the educational pipeline, let alone those older workers who need to return for further training.

There is also broad-based agreement that citizens in today's society need to have a greater understanding of science in order to make informed political and personal judgments. "Scientific literacy" is useful not only on many jobs but in a wide range of daily experiences. Citizens need to make judgments about environmental and energy issues. They are constantly confronted with experiences that require some information about health and medical care. The subjects of nuclear warfare and weapons systems will remain central topics of controversy. Scientific education, then, is important for personal well-being and for citizens' political efficacy. Life in the "new information society" will be more intelligible to those whose education has exposed them to scientific principles and applications.

The educational system is being asked to resurrect standards and expectation levels of a previous era. At the same time, it must adapt to an awesome economic transformation, one whose shape and outcome remains uncertain. These demands for changes occur at a time of shrinking public resources and diminished federal leadership in education. Thus far, the response of education to the high technology boom remains profoundly unsatisfactory and uneven. The rise of new industrial forces without concomitant alterations in education and training threaten the survival of new industries themselves as well as the prospect of viable employment for individuals in the transformed economy.

3 Two Case Studies in Neglect: Schools in California's Silicon Valley and Along Boston's Route 128

It is ironic that the nation's two leading centers of high technology, the Boston area and Santa Clara Valley, are also leading contemporary examples of deterioration in the condition of public education. As one Silicon Valley educator put it, "education is in a dismantling mode." The bankruptcy of the San Jose Unified School District during 1983–84, the first district in the United States to go bankrupt since the early 1940s, was particularly incongruous since it is located in one of the country's most affluent metropolitan areas. In both regions, the upbeat ambiance found in the high technology firms contrasts sharply with the demoralized atmosphere enshrouding the public schools. A wide range of school programs, including those related to science and mathematics, have been subjected to cutbacks in recent years. As a result, the capacity of schools to prepare students with the broad and demanding academic education they need in order to cope with changing employment conditions has been undermined.

Substantial employment in the high technology sector is already a reality in California's Santa Clara Valley (Silicon Valley) and in the Boston suburbs north and west of the city. Since these two areas have the highest concentration of jobs in high technology nationally, it is useful to examine the directions and condition of public education in these "knowledge-intensive" centers. A study of education in the heartlands of computers and electronics gives us some recent evidence about the extent to which educational institutions respond to changes in economic environments. Since job opportunities in high technology in these particular regions are projected to be relatively plentiful, the issue raised earlier of whether high technology industry can actually provide jobs for people is not as troublesome as it is in other geographic regions.

Similarities in Setting and Development

There are a number of historical and economic similarities between Boston's Route 128 area and Silicon Valley. Both have had a long history in the field of electricity and electronics. The General Electric Company began in Lynn, Massachusetts, just outside Boston, with the merger of the Edison General Electric Company and the Thomas-Houston Electric Company. GenRad Corporation began as General Radio of Cambridge in 1915, the first American electronics manufacturer.[1] On the opposite coast, Palo Alto is considered the "birthplace of electronics" since the vacuum tube was developed there by Lee DeForest and his colleagues in 1912. They were employees of Federal Telegraph Company, a radio company in Palo Alto founded in 1909 with the aid of Stanford University's first president. Both Magnavox Company and Litton Industries were founded shortly thereafter by other employees of Federal Telegraph.[2]

The presence of the Massachusetts Institute of Technology and Stanford University in the two regions was instrumental in the development of the industrial electronics complexes that grew up. The two institutions have provided brainpower combined with an entrepreneurial ethos, and they have continually nurtured the growth of technologically sophisticated firms, including fledgling biotechnology companies. Both have encouraged faculty to found new companies (including venture capital firms), consult regularly with businesses, and pursue collaborative research projects. Stanford has at times provided financial support for new business ventures and created the Stanford Industrial Park in the 1950s, "a sort of built-in Route 128."[3]

Another crucial ingredient in the successful development of both areas has been the infusion of federal defense funds. Although a number of electronics and other science-based firms existed in the Boston area prior to 1940, it was the massive federal support for war-related research during the early 1940s that provided the major impetus for the development of new technologies in the Boston region. A large number of new companies formed soon after the end of the war, many of them spin-offs from research laboratories at MIT, "the undisputed fountainhead of high technology firms in the area."[4] Route 128, the famed circumferential highway around Boston (the "Golden Horseshoe") built in the late 1940s and early 1950s, eventually became home for hundreds of high technology companies. Later, another circumferential highway, Route 495, was constructed west of Route 128, which also became a site for many such firms. Similarly, the industrial electronics complex around Stanford received a significant boost from federal contracts after World War II, urged on by Frederick Terman, dean of Stanford's School of Engineering, who had spent the war years at Harvard's Radio Research Laboratory working on radar countermeasures. Post-war military electronics contracts allowed Stanford scientists to develop new corporations, such as Watkins-Johnson and Applied Technology.

Federal contracts for the aircraft and missile industries have been of continuing import to both regions for decades, and federal procurement of semiconductors in the 1960s was crucial for the development of that industry. California and Massachusetts lead all other states in military research and development money. About one-third of the Massachusetts high technology work force was involved in government-funded military and aerospace research and manufacturing in 1983, funded by $6.3 billion in prime defense contracts. An estimated 16 percent of economic growth in Massachusetts between 1981 and 1988 will be accounted for by defense spending. California, with 10 percent of the nation's population, now garners about 18 percent of the U.S. defense expenditures. In 1983, $3.9 billion in prime military contracts flowed into Silicon Valley firms.[5]

In addition to the fact that these two areas share a long history in electronics, the presence of two leading academic institutions who aggressively foster industrial development, and decades of experience as beneficiaries of military contracts, they share other features as well. Both have benefited from the availability of large amounts of venture capital, unleashed by changes in the federal capital gains tax in 1978, which is another essential ingredient for the development of high technology complexes. Although the Silicon Valley area now has more venture capital available to entrepreneurs than the Boston region, both far outstrip other developing high technology centers in the country that lack a long tradition of technically sophisticated business formation.[6] Both remain centers of new company start-ups as a result of plentiful capital, proximity to new graduates emerging from Stanford and MIT, and the advantages that accrue as the result of the concentration of so many similar firms locating in one place.

These "agglomeration" effects mean that there are numerous support services (e.g., law firms, educational programs, investment analysts, real estate agents, headhunters, and management consultants specializing in high technology) and other employment opportunities available within one geographic area. Business analysts point out that the existence of this "high tech infrastructure" helps retain corporate headquarters in the two areas even though manufacturing units of the companies are located elsewhere. The network of services helps small companies survive during recessions and, most importantly, aids in the formation of start-up companies. Indeed, it is the continued growth of new firms and new technologies that give these regions their economic vitality. It is estimated, for example, that 200 new companies are formed each year in Silicon Valley.[7] The future economic health of these economic areas depends not so much on retaining older high technology firms whose product cycles are maturing, which means production has now become standardized and can be moved to other regions, but on the creation and expansion of new corporations.

The high technology industrial atmosphere of both regions gives them a sense of economic exuberance and optimism. Although both areas slumped economically during the early and mid-1970s, a period of reductions in military contracts as well as a recession, and suffered again in the recession of the early 1980s, their economies were far stronger than those where traditional "smokestack industries" were concentrated. And the companies in both complexes rebounded strongly in 1983 and 1984, with unemployment falling below 5 percent, when the period of national economic recovery began. Despite another dip in the semiconductor and computer manufacturing work force in 1985, employment projections in the two places are optimistic, and there is a general sense that the future belongs to these innovative firms. Finally, both regions rank high in what is called "quality of life," a factor that is seen as important in recruitment of managers and engineers. Both areas boast accessibility to a rich cultural environment and extensive recreational opportunities, and both are viewed as places where technical professionals can easily mix with many other like-minded people.

Regional Economic Variations

While the similarities between the two regions are important and obvious, observers find the differences more intriguing. There are obvious geographic differences. Santa Clara Valley is relatively self-contained, twenty-five miles in length and ten miles in width (from Palo Alto south to San Jose), bounded by mountains and foothills. While high technology companies have also spilled over into San Mateo County, southern Santa Clara County, and Santa Cruz County outside the Valley, the area has a more definable center than does the Route 128 region, which is more diffuse geographically and lacks physical boundaries. The economic histories of the region vary considerably prior to World War II. Northern Santa Clara County was an agricultural region until an industrial transformation began in the 1940s. In 1940, the county was ranked as one of the most productive fruit and vegetable producers in the country.[8] Its creation as a manufacturing center was a self-conscious effort on the part of Frederick Terman at Stanford, who envisioned and fashioned "a community of technical scholars" engaged in a complex of science-based industries around the university.

By contrast, the Boston area has had a long manufacturing history in textiles, machine tools, leather goods, refined scientific instruments (even in the nineteenth century), and was involved in the early manufacture of rubber and automobiles.[9] As labor economists have pointed out, the periodic revitalizations of the regional economy have depended for almost two centuries on the growth of new, technologically innovative firms.[10] Electronics and computers were simply another stage in this historical progression. The state's residents have lived through periods of major manufacturing decline, particularly in leather goods and textiles, as companies moved to other locations in search of cheaper labor or declined in the face of foreign competition. The fact that the region has a history of fairly recent industrial flight has made the populace more cynical about the long-range commitment of high technology companies to remain in Massachusetts. This cynicism is refueled periodically by the continued threats of high technology firms that they will expand elsewhere if state legislation unfavorable to their interests—such as laws requiring advance notice of a plant closing—is passed.

There are other economic differences as well. The high technology sector is a dominant manufacturing employer in Silicon Valley, accounting for 25 to 33 percent of the total work force, whereas high technology industry employs less than 10 percent of all workers in the Boston area.[11] In 1982, approximately 1,000 high technology manufacturing firms in Santa Clara County employed nearly 200,000 workers, and an additional 1,000 service enterprises (e.g., research and development, computer services) and distribution businesses employed another 32,000.[12] The employment figures grew during the period of economic recovery in the mid-1980s (a recovery fueled in part by increases in military contracts). Between October 1983 and October 1984, for example, nearly 14,000 new jobs were created in electronics firms in the Valley. Strictly comparable figures for the Boston area are difficult to calculate since data on high technology employment is usually given on a statewide basis. According to the Massachusetts Division of Employment Security, approximately 243,000 workers in Massachusetts were employed in 1983 by over 700 manufacturing firms in the high technology sector, and another 850 business establishments employed 23,000 workers in computer and data processing services and in noncommercial education and scientific research, most of them in the Route 128–495 area. As in Silicon Valley, employment in these firms grew substantially during the economic boom in late 1983 and 1984 so that by 1985, 276,000 people worked in high technology corporations.

Although there is clearly a higher concentration of high technology workers and firms in Silicon Valley than in the Boston area, Massachusetts as a state has a higher proportion of its manufacturing work force (35 percent in 1980; 40 percent in 1985) engaged in that sector than does California (30 percent in 1980) and ranks first among the ten largest industrial states on that index. Using definitions and data of Massachusetts agencies, approximately 9 percent of all employment in Massachusetts is in high technology manufacturing compared to 6 percent in California.[13] (Table 3.1) Thus, the high technology industry probably exerts more economic and political influence at the state level in Massachusetts than it does in California, but it does not permeate Boston area culture and politics with the same force that is found in Santa Clara Valley. The Boston area is a center of government, finance, insurance, hospitals and medical research, higher education, and tourism. While high technology manufacturing is the largest and most dynamic goods-producing sector in the area, it does not necessarily dominate the business environment as it does in Silicon Valley.

Silicon Valley companies and Route 128 firms have only recently had a fair degree of direct competition with one another because they have specialized in different products. The integrated circuit or semiconductor industry, the source of the Silicon Valley name since the circuits are etched onto silicon chips, was the principal source of high technology development in Santa Clara County. Twenty percent of the international market in semiconductors is accounted for by Silicon Valley firms.[14] Computer components, such as floppy disks, have also been a mainstay of the electronics economy, though the computer industry and aerospace and missiles were important forces early in the Valley's development. The growth of high technology in Massachusetts centered on radar and computers,

TABLE 3.1   Employment in High Technology Manufacturing Industries in the Ten Largest Industrial States, 1980

STATE	ALL INSURED EMPLOYMENT (IN 000's)	HIGH TECHNOLOGY EMPLOYMENT		
		Number (in 000's)	Percent of Manufacturing	Rank
Massachusetts	2,595.7	235.0	34.8	1
California	10,104.3	601.2	29.9	2
Florida	3,620.8	107.1	23.6	5
Illinois	4,692.9	237.0	19.1	6
Michigan	3,291.6	80.9	8.2	10
New Jersey	3,008.9	184.8	23.7	4
New York	7,113.6	374.5	25.7	3
Ohio	4,119.2	151.4	12.0	9
Pennsylvania	4,621.2	213.8	16.0	7
Texas	5,583.1	157.8	15.0	8

SOURCE: Massachusets Division of Employment Security, *High Technology Employment: Massachusetts and Selected States 1975–1981*, Boston, 1982.

especially minicomputers. The semiconductor industry has been virtually nonexistent there although some companies now produce chips for internal use. Over 50,000 jobs now exist in the Massachusetts minicomputer industry, and 70 percent of the minicomputer firms are headquartered in New England. However, both areas, particularly Silicon Valley, have developed diversified high technology companies that now include a computer software industry, peripherals, and biotechnology firms as well as businesses manufacturing a host of other products.[15]

Life-style Differences

Much has been made of the fast-paced, risk-oriented Silicon Valley culture. The famed Silicon Valley management style stresses informality, innovation, a minimum of bureaucratic redtape, and a downplaying of status differences. By contrast, the Route 128 area is viewed as being more traditional, even downright stodgy, in corporate and community life-styles. Some of the stereotypes have a factual basis. Venture capitalists in Silicon Valley are more willing to invest in new firms, and entrepreneurs seem open to taking more risks. As a result, more start-up firms are begun in the Valley than in the Boston area. One venture capitalist who operates in both areas puts it this way:

For every example of a successful start-up in Route 128 [today], there are five stories in the Bay Area.... On the East Coast people form a new business in a hush-hush way, working at their jobs during the day and putting together a business plan at night, which they circulate to the venture capital community hoping word does not get back to their employer. In California, entrepreneurs are more inclined to leave their employers and then go out and write a business plan and start raising money. Their attitude is: "Even if I don't succeed, I can always get another job."[16]

Silicon Valley engineers and managers have the reputation of being "workaholics working in a strike-it-rich environment," where "the accepted wisdom is that Silicon Valley has created more millionaires per square mile than any other manufacturing area on earth."[17] One report claims there are more than 3,000 millionaires in Santa Clara County while another puts it at 15,000.[18] Employees in high technology firms have a reputation for a high rate of drug usage and broken marriages. In fact, Santa Clara County has the highest divorce rate of any area in California and one of the highest in the nation.[19] Boston Computer Society president, Jonathan Rotenberg, claims that the image of California is that "everybody sits around all day with their guacamole and Quaaludes. In New England you still have, to some extent, your Yankee undertones."[20]

Like most stereotypes, these are exaggerated, but it is true that the Boston area high tech scene is more sedate. Rumors of drug use and high divorce rates are less common (indeed, Massachusetts has one of the lowest divorce rates of any state in the country) and wealth is not flaunted to the same degree. William Thurston, the chief executive officer of GenRad Corporation in Massachusetts, which is the oldest and one of the largest of the area's high technology firms, observed that "New Englanders seem to stay near their homes and visit their parents for Christmas and Thanksgiving. Their roots are deeper."[21] A survey of compensation strategies of forty-one high technology companies in the Boston area found that top executives spurned first-class airline flights, chauffeurs, corporate jets, and paid-up country club dues, preferring stock options and company-paid personal financial planning assistance.[22] Still, the Boston area high tech scene is sometimes characterized in terms familiar to those in Silicon Valley—engineers and programmers in beards, T-shirts, and jeans; a preference for expensive foreign cars; and long work hours by those engaged in the cutting edge of technology. Tracy Kidder's description in *The Soul of a New Machine* of the high pressured, round-the-clock efforts of young engineers developing a new minicomputer at Data General in Massachusetts could equally well have described similar teams in Silicon Valley.[23]

Demographic Profiles

There are some important demographic differences between Santa Clara County and the Boston metropolitan area. The Silicon Valley communities are wealthier, faster growing, and ethnically more diverse than those in Boston and its surrounding towns. Santa Clara County has boomed in population in recent decades: its population increased by 26 percent to 1.3 million between 1970 and 1980. San Jose, its major city, had a 1982 population of 672,000, larger than that of Boston, which slipped by 12 percent to 563,000 between 1970 and 1980. (The city's population, however, is now growing once again.) The county was one of the nation's fastest growing during the 1970s. The Boston area, by contrast, declined by almost 5 percent during the decade to about 2.8 million.

Santa Clara County has a much larger Hispanic population than the Boston area, 17.5 percent versus only 2.4 percent of the residents. Twenty-two percent of the school children in Santa Clara County in 1983 were Hispanic. Its Asian population, which includes many Vietnamese, is also more substantial, 7.8 percent compared to less than 1 percent in the Boston area. The percentage of blacks in both metropolitan areas is quite low, 3.3 percent in Santa Clara County and 5.8 percent in the Boston metropolitan area. (Table 3.2) The ethnic composition of the student population in Santa Clara County has been undergoing dramatic change. Since 1971, the Asian population in the county has more than tripled and the Hispanic population has nearly doubled in size. Entering elementary classes are already more than 50 percent minority in the San Jose Unified School District, a phenomenon that will become county-wide by 1990.[24]

Differences in family income are striking. The population of the city of Boston is much poorer than that of San Jose: Only 8 percent of the San Jose population was classified as poor in the 1980 U.S. census while 20 percent of the Boston population fell into the poverty category. The 1980 median family income in Boston was $16,253 while that of San Jose was $22,886. Indeed, the San Jose metropolitan area has one of the highest median family incomes of any metropolitan area in the country. Santa Clara County white fami-

TABLE 3.2 Racial and Ethnic Composition of the Population of San Jose SMSA/Santa Clara County, Boston SMSA, City of San Jose, and City of Boston, 1980

	HISPANIC	BLACK	ASIAN	TOTAL POPULATION
San Jose SMSA (Santa Clara County)	17.5%	3.3%	7.8%	1,295,071
Boston SMSA	2.4	5.8	.01	2,763,357
City of San Jose	20.9	4.3	7.6	671,800(1982)
City of Boston	6.4	22.4	2.7	563,000

SOURCE: U.S. Bureau of the Census, 1980.

lies had a median income of $30,966 in 1980 while white families in metropolitan Boston averaged $23,514. Silicon Valley's minority groups fare substantially better in median family income than do minorities in the Boston area: black families in Santa Clara County averaged $22,842 in 1980 compared to $12,944 in metropolitan Boston; Hispanic families made $22,066 while Boston area Hispanic households averaged a mere $11,390; and Asians made $29,428 compared with $19,833 made by their counterparts in Boston and its environs.[25]

The average annual wage of Santa Clara County workers ($20,536 in 1982) is the fifth highest of all U.S. metropolitan areas. However, median family incomes in Boston's high tech bedroom communities were equal to or higher than those in comparable suburbs in Silicon Valley. [26] And the unemployment rate in the Boston area and in Massachusetts was lower than that of Santa Clara County and California throughout the recession of the early 1980s. Both regions have a high cost of living, with the cost of housing in Santa Clara County and the Boston area being among the highest among metropolitan areas in the United States. To even the casual observer, however, a drive through the two areas shows that the California high technology center as a whole is considerably more affluent than its Massachusetts counterpart.

The Condition of Public Education

Both of these high technology regions have something else in common: Their public educational systems deteriorated in the late 1970s and 1980s. Data gathered for this study from interviews with school officials in seven Route 128 communities and eight Silicon Valley school districts and from a variety of other sources confirm that their local public schools have declined in enrollment and financial support and have experienced a deterioration of the teaching environment. Even the highly esteemed "lighthouse" school systems, such as Palo Alto and Newton, have demoralized faculties who feel their school programs have lost some of the innovation, vision, and excitement that have accounted for their national prominence. In the early 1980s, the downward slide was more pronounced in Silicon Valley schools than it was in the Boston area, but by 1985, the Route

128 school systems were approaching the crisis conditions that hit the California schools a few years earlier.

ENROLLMENTS

Enrollment drops have had a traumatic effect on school systems in both areas. Overall, Massachusetts school districts have been hit much harder by losses of students than have California school systems, but some of the more affluent California districts, including some in Silicon Valley, have lost half their enrollment in the last decade. Between 1971 and 1981, public elementary and secondary school enrollment dropped 12.1 percent in California (close to the national average of 12.8 percent) but fell by a hefty 16.3 percent in Massachusetts. (Table 3.3) The Bay State closed 23 percent of its public school buildings in that same ten-year period, the second-largest reduction of any state in the nation. California, by contrast, had a 6.8 percent increase in public school buildings.[27] Projections of future enrollments show that California's school-age population will grow by 32.4 percent between 1985 and the year 2000 while the Massachusetts school-age population will increase by only 2.1 percent over that time period.[28] New England has historically had low fertility rates. The only area of the country with a lower fertility figure than that of Massachusetts currently is Washington, D.C. One demographer estimates that only half of the women in the New England cohorts born during the mid-to-late 1950s will have more than one child.[29]

TABLE 3.3   Percentage Changes in School Enrollments 1971–2000, United States, California, and Massachusetts

	1971–81*	1985–90	PROJECTED† 1990–2000	1985–2000
United States	−12.8%	+ 5.3%	+ 12.2%	+ 18.2%
California	−12.1	+ 10.7	+ 19.6	+ 32.4
Massachusetts	−16.3	−4.0	+ 6.4	+ 2.1

SOURCE:
*National Center for Education Statistics, *The Condition of Education, 1983 Edition*, statistics based on public elementary/secondary school enrollment.

†U.S. Department of Education, *Prospects for Financing Elementary/Secondary Education in the States Volume I*, 1982, statistics based on projected school-aged population.

Comparative data between Silicon Valley and the Route 128 area are difficult to construct since the Massachusetts electronics belt has less definite geographic boundaries, but the evidence available shows that the Boston area schools overall have suffered greater enrollment losses than their counterparts in Silicon Valley. For the purposes of this study, six communities with the highest concentrations of high technology firms in the Route 128 area were selected for study along with one other Route 128 city where many technical professionals reside. These seven municipalities averaged a 6.7 percent enrollment decline in 1981–82, slightly above the state average of 4.2 percent. During that same year, the school-age population in Santa Clara County fell 3.2 percent, with the peak loss year being 1978 when a 4-percent enrollment decline occurred. In 1982–83, enrollment across the county dropped only 1.8 percent.[30]

However, enrollment trends in Santa Clara County are uneven, with school districts in the southern part of the county (Morgan Hill, Gilroy) and on the east side of the city of San Jose showing a growth in student population while those in the western and northern sections of the county posting losses. Wealthier communities where prohibitive housing prices keep out young families have suffered the greatest drops in their school populations. Enrollment is expected to plummet by 30 percent between 1983 and 1988 in the Fremont Union High School District, which covers Sunnyvale, Cupertino, and a small pocket of San Jose. Even greater reductions in enrollment have been recorded in the area's wealthiest towns, such as Los Altos and Palo Alto. The elementary schools in exclusive Los Gatos and Saratoga had drops of 35 percent and 38 percent respectively in their student populations between 1970 and 1980.[31] These declines are comparable to the massive reductions in the student cohorts in the Route 128 communities.

EDUCATIONAL FINANCE:
THE LEGACY OF PROPOSITIONS 13 AND 2 ½

A second dramatic influence on the public schools in these two regions is decreased financial support for public education. Here, school systems in California have suffered more than those in Massachusetts, but both have been jolted by property tax cutting schemes passed by the voters as well as a variety of other budget reduction policies. Even though enrollments have decreased, school costs have remained high so that budget reductions are painful. While it seems plausible that school expenditures should automatically drop with reductions in enrollments, only about 10 percent of the average cost per student is saved when a student position is lost if drops in enrollment are evenly distributed across classes and schools in a district.[32] School systems in both states had been subjected to fiscal austerity measures even before the passage of Proposition 13 in California in 1978 and Proposition 2½ in Massachusetts in 1980, the ballot initiatives that placed ceilings on local property tax rates.[33] Because of the reduction in local revenues brought about by Proposition 13, whose impact was initially cushioned by a $5-billion surplus of state funds, about three-fourths of California's revenue for public schools now comes from the state ("the big school board in Sacramento") so that any discussion of school finance there must now center on state politics. Massachusetts' school districts receive only 40 percent of their school budgets from the state. In both states, there has been a significant drop in the number of candidates for school boards since the budget reduction measures were passed.

The school budget situation in California has been particularly grim: Budget reductions caused California's public schools to be ranked well below national averages in spending per pupil in the early 1980s and in total expenditures as a percentage of personal income, a measure of the state's educational effort. (Table 3.4) School revenues in California dropped about 16 percent in real dollars between 1977 and 1982.[34] The San Jose Unified School District was declared bankrupt in 1983 and several other districts considered filing for bankruptcy. However, a significant reversal of policy occurred in 1983 when a strong coalition of groups in the state was able to push increased funding for the public schools through the state legislature and to win gubernatorial approval. As a result, some cuts are being restored so that California schools now have funding levels comparable to the national average. Three billion additional tax dollars were allocated to the California public schools between 1984 and 1986, but the amounts were only enough to begin the rebuilding process. The upturn in funding must be sustained for several years more if schools are to resume previous levels of services.

The long-range outlook for adequate funding is still worrisome, particularly in view of the fact that California schools will have growing enrollments and increasing numbers of students who have limited proficiency in the English language, which will necessitate

TABLE 3.4   Rankings of California and Massachusetts
on Selected Educational Indicators

	CALIFORNIA STATE RANKING	MASSACHUSETTS STATE RANKING	NATIONAL
Estimated current expenditures for public elementary and secondary schools per pupil in average daily attendance, 1984–85*	$3291 (26th)	$3889 (13th)	$3429
Total current expenditures for public elementary and secondary schools in 1981–82 as a percent of personal income, 1981†	3.72 (44th)	3.98 (38th)	4.20
Pupils enrolled per teacher in public elementary and secondary schools, fall, 1982†	23.34 (2nd)	17.79 (27th)	18.54
Estimated average salaries of public school teachers, 1984–85*	$26,300 (7th)	$24,110 (19th)	$23,546

SOURCE:
*National Education Association, *Estimates of School Statistics, 1984–85.*
†National Education Association, *Rankings of the States, 1982.*

additional school services. More than 25 percent of California's public school students are of Hispanic origin and there is a growing percentage of newly arrived Southeast Asians.[35] At Independence High School in San Jose, for example, 34 percent of the school's 4,400 students were born in a foreign country and, among those, forty-four different languages are spoken.

Massachusetts has traditionally given high priority to its public schools. In 1981–82, the state ranked fourth in the nation in expenditures per pupil and fourteenth among the states in expenditures as a percent of personal income. During the latter half of the 1970s, Massachusetts' percentage increase in real expenditures per pupil was more than double that of California. But since the passage of Proposition 2½, Massachusetts' rankings among the states have dropped. Estimates from the National Education Association placed the state at thirteenth in expenditures per pupil in 1984–85 (twenty-fourth in 1983–84). The Bay State spends $1,300 less per pupil than New York State and $600 less than Connecticut, although these neighboring states have similar economies and costs of living.[36]

The school day and year in California are comparatively short, in part because of efforts to save money. Until implementation of new state regulations adding days onto the school year in 1984, California students who attended public schools in the state from kindergarten through twelfth grade received 1⅓ fewer years of instructional time compared to other students in the United States as a result of the truncated school day and year.[37] Even if the new regulations are put fully into effect by 1986–87 as planned, California students will still lag behind national averages in instructional minutes. The pupil-teacher ratio in California is the second highest in the country. The Massachusetts ratio was the lowest in the U.S. in 1981–82, but it rose substantially beginning in the fall of 1982 so that it is now about the same as the national average. (Table 3.4) These two states were the only ones in the country between 1972–73 and 1982–83 to register an increase in pupil-teacher ratios in public elementary schools, a reflection of the impacts of Propositions 13 and 2½. Teachers' salaries in California are comparatively high, averaging $26,300 in 1984–85, which placed them as the seventh highest in the nation, considerably above the national average of $23,546. Even so, the purchasing power of teachers' salaries declined by 15 percent in that state between 1972–73 and 1982–83. Massachusetts ranked nineteenth among the states in average teachers' salary, $24,110, in 1984–85.[38]

Massachusetts school budget cuts have taken their toll. During 1981–82, school systems' municipal budgets were cut an average of 5.5 percent, which, with inflation taken into account, amounted to more than a 10 percent budget reduction. Fifty-three districts had their municipal school budgets cut by more than 10 percent before inflation. Approximately 8,000 teachers were laid off as declining enrollment and the impact of Proposition 2½ led to a 12-percent average cut in professional staff. (Half of these layoffs were due to declining enrollment and half to the effects of Proposition 2½.) An additional 1,800 teachers were laid off in 1982–83 and hundreds more lost their jobs in 1983–84. Between 1972–73 and 1982–83, the proportion of Massachusetts elementary public school teachers dropped by one-fourth, the largest percentage decline in the country.[39]

In 1981–82, 163 school buildings in Massachusetts were closed as a direct result of 2½ alone, and half the local school districts established user fees for such services as athletics and instumental music lessons. Expenditures for new equipment declined 36 percent while textbook purchases dropped by 30 percent in that same year.[40] At least sixty of the state's school superintendents announced their retirements or resignations during 1981–82, a dramatic increase from the average of fifteen in previous years.[41] Budget reductions since 1980 have exacerbated the already wide variations in per pupil expenditures between school districts, with some districts spending as much as $5,000 per pupil and others as little as $1,600 in 1984. A recent federal study forecast favorable long-range funding prospects for Massachusetts, based on its traditional high effort for education and lack of significant anticipated growth in enrollment in the next decade and a half.[42] However, in the future, public schools will be competing for the public dollar with the health and service needs of an aging population.

Schools in Silicon Valley

The twin effects of declining enrollment and budget reductions have created a profoundly transformed environment within which public schools must operate. The effects are shockingly apparent in the public schools of affluent Santa Clara County. All but four of the county's thirty-three school districts spend less per pupil than the statewide and national averages.[43] A 1981 study for the Santa Clara County Office of Education found that the money available for traditional school programs dropped in real dollars from 1974 to 1981, with increases being registered only for categorical programs for the poor and handicapped.[44] The fact that the largest school district in the Valley, San Jose Unified with 32,000 students, was subsequently declared bankrupt highlights the contradiction between private wealth and public austerity. An in-depth investigation of Santa Clara County by a team of *Washington Post* reporters in 1983 led them to conclude that "probably nowhere in America is there a greater disparity between private affluence and declining public services."[45] The financial condition of the schools has improved somewhat since 1983, but the rebuilding process is a slow one.

Many of the school districts in Silicon Valley have laid off all their younger teachers as a result of budget cuts and/or declining enrollment. Classes in Silicon Valley are large, (as many as thirty-eight students) considerably bigger than those in the Route 128 area, especially at the high school level. Numerous school districts in California eliminated the sixth period of the senior high school day following passage of Proposition 13, a step taken by the San Jose Unified School District for juniors, seniors, and middle school students in 1979. And the sixth period became optional for students in the East Side Union High School District in San Jose. The elementary school day was also shortened in some districts. During the first half of the 1980s, seventh-grade students in Santa Clara County averaged only 265 minutes per day in class (4.4 hours), compared to the statewide average of 307 minutes and a national average of 332 minutes. Between 1979 and 1985 eighth graders enrolled in San Jose Unified spent 250 minutes per day in class (50 minutes less than before Proposition 13) compared to an average of 333 minutes a day nationally.[46]

Some additional state funding to extend the school day and school year was made available beginning in the fall of 1984, which provided a start in alleviating the problem. San Jose Unified, for example, was able to restore a required six-period day to high school juniors and middle school students during the 1984–85 school year. Further, almost all of the county's school districts are in the process of adding several days to the school year in an attempt to reach the national average of 180 days. Efforts to bring the country and state schools up to or near national averages cost a good deal of money. In the San Jose Unified School District alone, it will cost $3.7 million a year to resume a 300-minute day for all students in the district.

Summer school classes have been curtailed in Silicon Valley schools as they have throughout California since the passage of Proposition 13. The school textbook situation is particularly severe. Textbooks used to be replaced every six years, but ten-year-old texts are now common in Santa Clara County. One elementary school science text still used widely in the state was published in 1965 and still talks about "how man will someday walk on the moon."[47]

Students have become used to sharing texts, and teachers commonly reproduce instructional materials when texts are unavailable.

Guidance counselors and librarians have disappeared from many public schools in Santa Clara County, replaced sometimes by parent volunteers. San Jose Unified, for example, eliminated all junior and senior high school guidance counselors and half the librarians during the worst years of retrenchment in the early 1980s. Fremont Union High School District, whose boundaries include the electronics heartlands of Cupertino and Sunnyvale, also laid off all librarians and counselors. East Side Union High School District in San Jose cut its counseling staff by half. And in 1984, Cupertino (the headquarters of Apple Computer and many other innovative and thriving electronics firms) had to lay off a number of its counselors and school librarians when a special tax measure failed to garner the two-thirds voter support needed for the proposal's passage. Again, the 1983 state legislation targeted more money for textbooks, summer school, and guidance counselors, but whether the amounts allocated are sufficient restorations to the budget remains unclear.

The number of small advanced and remedial classes has been reduced to save money. The Mountain View School District, for example, eliminated its entire remedial reading program and a quarter of its courses for gifted children in 1982 in order to balance the school budget. Coaches who are also faculty members have been increasingly replaced throughout the Valley and state by "walk-on coaches" who are hired only for their coaching activities. San Jose Unified had to cut junior and senior high school athletic programs in half. Students in a number of districts were paying fees for participation in sports and in other extracurricular activities until a 1984 California Supreme Court decision declared such fee impositions unconstitutional. One Valley school district began charging students a fee to ride school buses in 1984. Elementary school classes in art, choral music, and instrumental music were among the first casualties of Proposition 13 in most school districts in the Valley. In San Jose Unified all elementary instrumental music was eliminated for several years and only one high school now has a full band program. Marching bands were disbanded in some high schools in the Valley. Staff development programs for teachers, including sabbaticals, have been greatly cut back as well in many of the county's school systems. The need to hire instructional staff to teach the rapidly growing number of non-English-speaking students also has placed a strain on school budgets.

The decline of public education was illustrated in a *San Jose Mercury News* profile of Willow Glen High School, a school with a demographically representative student population for the city of San Jose.[48] In the years following the passage of Proposition 13, the high school eliminated the sixth period for juniors and seniors, laid off teachers, and assigned some teachers to classes they were not qualified to teach. In one instance, a veteran teacher whose last laboratory experience had taken place twenty years earlier was assigned to teach chemistry. Small English classes were abolished as were allocations to pay department chairs to coordinate curriculum and budget. All guidance counselors were laid off, resulting in a drop of $300,000 in scholarship awards in one year alone to the school's graduates, 40 percent of whom go on to college. The amount budgeted for students' books and supplies was reduced from $40 per pupil in 1977–78 to $31 ($19 considering inflation) in 1982–83. Groups of six to eight classes of students had to share a set of books, requiring students to check books out overnight in order for them to study for tests in some classes since students did not have individual copies. Industrial arts teachers lack adequate supplies, such as wood and metal, for shop projects.

The problems are not confined to the city of San Jose, whose difficulties have received the greatest attention. The effects of fiscal austerity are visible even in the schools of highly affluent communities as well. Saratoga, for example, is a wealthy residential community whose median family income in 1980 ($41,143) was exceeded in the county only by Los Altos Hills. The average selling price of a home there in 1983 was $230,000. Students at the high school rank high in academic achievement, yet the school lacks money to hire the three of four more teachers that would be required if the school were to add a needed seventh period to the high school day. During the spring of 1983, almost half the school's custodial staff (2½ people) were laid off, which makes the completion of routine maintenance tasks problematic, and school administrators were cut back to an eleven-month salary. Teachers have been assigned to instruct classes in subjects outside of their true areas of specialization. Faculty hired since the early 1970s have been laid off, in part because of declining enrollment, so that teachers' average age is in the forties. The school principal also had to send a letter of solicitation to parents asking for money for textbooks, with the goal of raising $15,000.

Along with several other affluent districts, parents in the Los Gatos–Saratoga High School District have established a foundation to raise money for their public schools. Parents in the less affluent East Side Union High School District and its feeder elementary school districts have also established one (with substantial corporate support) to raise funds for sports programs and classroom projects. Across the state, more than 300 school districts have formed these private foundations to augment insufficient school funds.

Cutbacks in mathematics and science programs have also occurred in Silicon Valley schools, although a few programs are being restored as a result of increases in statewide educational funding since 1983. The position of the science curriculum consultant in the Santa Clara County Office of Education was abolished in the wave of budget reductions after Proposition 13's passage. High school science and mathematics classes are often large (thirty-five to thirty-eight students) even in wealthy districts. Some of the high schools have not been able to purchase new textbooks in such fast-changing subjects as biology, physics, and chemistry in ten years. A high school chemistry text can cost more than $20, an expense many of the Santa Clara County schools cannot handle. Palo Alto High School retired a set of 1971 physics textbooks in 1985. At one high school in south San Jose, a school's parents' club tried to raise $18,000 to purchase new texts. At another high school in the city, where the science department's budget was slashed from $8,900 in 1978 to $1,300 in 1983, the Parent-Teacher Association worked to obtain chemicals for the chemistry department and dissecting materials for the biology department by approaching businesses for donations of the needed items.

The Mountain View–Los Altos High School District had to rely on donations from its privately funded education foundation to remodel and expand two science classrooms. Other foundations, which usually rely on parent contributions, have provided the money for essential science equipment. The fact that fundamental teaching materials and supplies for science courses are unavailable through public funding shows how far the deterioration of school programs has progressed in a region billed as the world's most innovative scientific community.

Route 128 Schools

Most Massachusetts public schools have not yet experienced the same degree of austerity that has characterized California and Silicon Valley schools. However, the Bay State schools are clearly repeating the devastating experience of the California schools after the passage of Proposition 13. Although schools in urban areas have been hardest hit by the impact of Proposition 2½, more affluent districts, such as those on Route 128, have suffered as well. The long-range financial picture in those systems is bleak. These districts saved large sums by closing schools in the 1970s and 1980s but, now faced with the prospect of slightly rising enrollments at the elementary level (which will be felt at the secondary level in the 1990s), they cannot continue to "buy time" financially by closing more schools. Although inflation is running between 4 and 6 percent, Proposition 2½ allows municipalities to increase revenues only 2.5 percent a year. This means that new rounds of cuts will have to be implemented every year as revenues fall short of cost increases. New state aid will go primarily to the poorer communities. Thus, the Route 128 school districts, which have traditionally had strong reputations, face a dismal fiscal future.

In the year following passage of Proposition 2½, Massachusetts public schools experienced painful budget reductions. During 1981–82, the budgets for extracurricular athletic and student body activities statewide declined by 21 percent and 29 percent respectively. There was a large reduction in librarians and audiovisual specialists (24 percent), reading teachers (22 percent), and guidance counselors (18 percent). Layoffs of art, drama, music, and foreign language teachers occurred at above-average rates. The pupil-teacher ratio increased in 1982 in 273 out of 379 operating districts. Cuts remained deep in many communities in the years subsequent to 1981–82. For example, thirty-six districts had more than a 10-percent budget reduction (before inflation) during the second year of Proposition 2½'s implementation.[49] Some districts, particularly those in urban working-class cities and towns, have had their educational programs profoundly harmed by budget caps and cuts.

Suburban Route 128 school districts have not suffered as much as these more urban areas, but have been forced to implement painful cuts. Burlington, for example, one of the major high technology centers on Route 128, laid off 15 percent of its professional staff in 1981–82, including sixty-six teachers (70 percent of whom lost their job directly as a result of budget cuts, not declining enrollment). Cuts were felt especially in special education, guidance, and alternative education as well as in art and social studies courses at the elementary level.[50] Another major high technology municipality on Route 128, Waltham, experienced significant budget reductions as well. Nineteen percent of the professional staff in the district were laid off during 1981–82. Overall, of the seven Route 128 communities studied in depth for this research, all but one (Bedford) experienced an absolute cut in their school budget in 1981–82. Professional staff reductions ranged from 8 percent to 19 percent, with an average of 12 percent for all towns, the same as the statewide percentage.[51]

While cuts in high school programs have been most pronounced in athletics, industrial arts, music, and assorted activities and clubs, there have also been financial reductions affecting the teaching of science and mathematics in these Route 128 school districts. Six of the eight science departments and half of the mathematics departments studied lost staff positions during 1981–82. (Math teachers are less likely than science teachers to be laid off because of increased student demand for computer electives and small remedial courses.) Eleven of sixteen mathematics and science departments had their budgets for equipment, supplies, and texts cut or level-funded during 1981–82. The chairs, particularly in the science departments, felt that these financial reductions were especially harmful because there has been such dramatic inflation in the costs of chemicals, glassware, and textbooks. One math chair commented that calculators were needed, supplies of paper were running out, and he "hoped that ditto fluid will last until the end of the year." Several stressed the need for more microscopes and balances. One science chair, who was "pretty mad" about the equipment situation in his department, pointed out that physics had not had any new equipment in thirteen years. Six of the eight high schools were forced to use mathematics and science textbooks for seven or eight years instead of the previous standard of five years. Two of the schools had lost funding for a faculty person to serve as coordinator of the school's math team. All of the schools believed that financial considerations had prevented them from acquiring the computer hardware and software necessary to accommodate student demand.

In addition to these reductions, one school lost a remedial mathematics laboratory and another lost the double laboratory period used in some science courses in the 1981–82 school year. Three schools had teacher sabbaticals abolished or in-service training programs eliminated. Four science departments and one mathematics department experienced an increase in class size or required teachers to take on an additional class. One school had to eliminate some of its ability groupings. Another cut eight courses from its science curriculum. Statewide, a number of towns eliminated the position of science coordinator, particularly at the elementary school level. Field trips to Boston's Museum of Science dropped by 25 percent during this first year of implementation of Proposition 2½. Overall, two-thirds of the mathematics and science teachers surveyed in the eight high schools claimed that cuts or constraints in the school budget had hurt courses in their discipline. They singled out lack of money for equipment, supplies, texts, and computers as their major concern followed by the growth in class size.

These obstacles to good teaching have not disappeared with time. Indeed, they have grown worse. A 1983 survey of secondary science supervisors in Massachusetts found similar deficiencies, and concluded that the state's formerly vigorous science education program was "alive but ailing."[52] Inadequacies in materials and texts were especially acute in urban public schools. Almost two-thirds of the public schools reported that their science budgets had been cut between 1978 and 1983 with the average reduction amounting to 34 percent. Nearly three-fifths of the science supervisors said that reduced funds for supplies and materials often resulted in fewer laboratory activities, the elimination of some experiments, and the omission of certain study topics. Approximately two-fifths said they were teaching with outdated texts. It should be noted, however, that class sizes are still quite reasonable, around twenty-four students per class, smaller than those in California.

Several of the Route 128 school districts, following in the footsteps of their counterparts in California, have recently begun to create private foundations to support the public schools. The hope is that such foundations will be able to purchase science equipment and fund other concrete needs. The formation of these organizations in Massachusetts illustrates the belief of parents' groups that public funding for schools will be inadequate for a long time to come.

Industrial Arts and Vocational Education

The condition of industrial arts courses and vocational education programs in both California and Massachusetts is also problematic. The educators interviewed in this field in the early 1980s were the most pessimistic of any group surveyed in the study of the two regions. Budget reductions following the passage of Propositions 13 and 2½ hit these programs especially hard. A survey by the California Teachers Association found that one out of every four school districts in the state had cut back on vocational education and industrial arts courses as a result of budget reductions associated with Proposition 13.[53] The fact that many school districts shortened the school day by eliminating or making optional the sixth period forced reductions in industrial arts electives typically offered in that period. There was little money allocated to purchase new equipment in Silicon Valley school vocational programs for a period of years after 1978. At the Regional Vocational Center in San Jose, a highly regarded school serving six school districts, which offers programs in such high demand fields as computer operating and electronics, the capital outlay budget was slashed from approximately $85,000 a year to almost nothing. The budget picture brightened after 1983 increases in the state's school outlays so that the RVC could purchase essential new equipment and carry out some of the repairs and refurbishing needed in its facility. Equipment and repair budgets are still inadequate in other parts of the county, but the situation is not as severe as it was.

Vocational educators in the regional occupational programs in Silicon Valley and the rest of the state have had to be resourceful in order to survive in an era of fiscal austerity. Administrative staffs have been pared sharply, class sizes have increased, school-provided buses have been dropped in favor of regular public transportation in some cases, and company sites have been used increasingly for instruction when equipment was too expensive to purchase. The worst of the cuts are over for the moment and the system is being reorganized, rebuilt, and tied in more efficiently with other sectors of the educational system. A new problem has emerged, however. Vocational enrollments have dropped significantly since the passage of more stringent academic graduation requirements in California in 1983. Students find it difficult to complete both their academic coursework, taken in their home district comprehensive high schools, and the coursework in the vocational school settings.

Interviews with occupational-education directors in the seven Route 128 school systems studied and with directors of two regional vocational high schools on Route 128 revealed that similar problems beset vocational and industrial arts programs in the Boston area. These programs have not yet rebounded from budget caps of the 1970s and the cuts resulting from Proposition 2½. Like programs in California, much of their budgets are tied up in equipment and supplies, which are inflating rapidly in cost. Three school systems had relatively good equipment and there were funds to maintain it. But five of the nine schools now have no money to replace equipment. "The budget cutting process has been devastating," claimed one occupational education coordinator who had seen his annual capital outlay budget slashed from $10,000 to zero. Proposition 2½ has had an especially negative impact on separate vocational schools that are located in urban areas.

Local tax cuts have been compounded in occupational programs by reductions in federal vocational funds under President Reagan. The U.S. Department of Education had a budget of $902 million for vocational and adult education in 1980, but that was down to $742 million by 1982. In 1985, the budget was back up only to $838 million, still short of the 1980 allocation. The federal appropriation for vocational education declined by one-third in real dollars during Ronald Reagan's first presidential term.[54] Prospects for increases in appropriations are dim. Yet federal funds have been particularly important in the development of new training programs in high technology fields. In 1980, one out of every two dollars for new vocational education programs came from the federal government.

The highly regarded twenty-seven regional vocational-technical high schools in Massachusetts experienced a real crisis in funding in the early 1980s. They derive their support from assessments of their member school districts, which, during the current funding reductions, are more reluctant to give money to a school that may take only a few of the district's students. There was no special provision made for the funding of regional schools in the Proposition 2½ legislation, and these schools got less state aid proportionally in

1981–82 than local school districts. The first programs to be cut were courses that served post-secondary students. After that, some programs for secondary students were eliminated and teachers laid off in order to finance pay raises for remaining teachers.

Capital outlay budgets for some of these Massachusetts regional vocational schools have been cut from $100,000 a year to almost nothing. As a result, certain programs are becoming obsolete. Four years after the passage of Proposition 2½, these schools still have no stable and adequate source of funding for capital outlay budgets to purchase equipment. Administrators spend large blocks of time on budgetary matters rather than on curriculum and administration, going from town meeting to town meeting pleading for adequate funding. A further problem that these schools face is increased competition with local high schools for students, a direct result of declining enrollment. In the past the showcase regional vocational-technical schools had waiting lists of prospective students, but increasingly some are engaged in an "outright war over bodies" with their member districts' high schools.

The Paradox

Thus, public schools in Silicon Valley and in the Boston area, beset by enrollment declines and budgetary cutbacks, have a reduced ability to provide students with the skills and knowledge they will need in a technologically sophisticated employment market. While many private corporations in the same communites grow and prosper as they create the "new information society," schoolteachers try to instruct with outdated texts and inadequate supplies. The contrast between private affluence and public poverty has been illustrated most vividly by the bankruptcy of the San Jose Unified School District in a county that reportedly has more millionaires than any in the world. Cutbacks and retrenchment schemes still preoccupy many school administrators as neighboring high technology managers plan innovative products and project optimistic growth curves. At a point when industry stands at the threshold of a new microelectronics era, schools look back to an earlier time as their golden age.

4 The Anatomy of Neglect: Trends in Student Achievement, Curricula, and the Teaching Profession

The development and advance of the "second industrial revolution" brought about by microelectronic technology in the 1970s occurred at a time when American school curricula became less stringent, expectations for performance slackened, achievement scores declined, and new entrants to the teaching profession became less qualified. These national trends in educational indicators have been apparent in both the Silicon Valley and Boston areas, the leading centers of scientific sophistication. The educational slide of California schools during the 1970s and early 1980s was more dramatic than that which occurred in Massachusetts, but the Bay State public schools are now experiencing many of the same problems that beset their Pacific Coast counterparts.

The two coastal regions have educational problems that mirror nationwide trends. But the variations between them provide contrasting case studies that show the diversity of school policies in the United States. Likewise, schools in Silicon Valley and in the Route 128 area, while reflecting state trends, still exhibit variations among themselves in curriculum, achievement, and issues affecting the teaching staff. This chapter looks at the general state of education on these three points but focuses especially on the teaching of mathematics, science, and computer courses, subjects that are becoming more essential as students prepare for the unpredictable and shifting occupational demands brought about by technological change.

Student Achievement: California and Massachusetts

Both California and Massachusetts have a relatively well-educated populace with California residents ranking sixth in the nation in median years of education among adults and Massachusetts ranking fifteenth.[1] California, however, has a higher percentage of high school dropouts than Massachusetts: In 1982, only 69 percent of California seventeen-to eighteen-year-olds completed their secondary education, placing them thirty-ninth among the states on that index, while 76 percent of Massachusetts youths did so, giving that state a ranking of twenty-third in the nation.[2] But a higher percentage of California school children were placed in gifted and talented programs (4 percent) compared to those in Massachusetts (1.2 percent).[3] New England students do a little more homework than the national average although the percentages are not impressive in absolute terms. California students have homework loads that are in line with the national average. In New England, 31 percent of the 1980 seniors in the *High School and Beyond* study had five or more hours of homework a week compared to 25 percent nationally and 24 percent in California. New England seniors have significantly lower rates of absenteeism and tardiness than their counterparts on the West Coast.

There is considerable evidence that students in Massachusetts have somewhat higher academic achievement levels than pupils in California. Variations by state in levels of achievement are difficult to assess because most studies group students by larger geographic regions and no uniform test except the National Assessment of Educational Progress, which thus far has not reported state breakdowns, is given to representative numbers of students in all states simultaneously. Thus, for purposes of this analysis, data on the Northeast or New England will be used in addition to the few state reports that are available. Comparisons of Scholastic Aptitude Test (SAT) scores between states are available, but they are not always useful since the proportion of seniors taking the test varies from state to state. Nationwid?, approximately one-third of all eligible seniors take the test, and 38 percent do so in California; a much larger pool, 63 percent, take the test in Massachusetts, a pool that includes many bilingual and special-needs students. As a result of this larger group of test takers, the SAT scores of Massachusetts students hover around the national average. (Table 4.1) It is noteworthy, however, that even with this less select and less affluent population of test-taking students, Massachusetts students' verbal score on the 1984 SAT was 8 points higher than that of California students. California students scored 9 points higher in the mathematics portion of the test. The SAT scores of the students in both states dropped during the 1960s and 1970s as did that of pupils across the country, although since 1982 the decline has been replaced by a modest upward trend, (except in California's verbal score, which has not increased) (Figure 4.1) The decline in scores during the 1970s was more precipitous in California than it was in Massachusetts.

There are other pieces of evidence which show that California student achievement ranks below that of Massachusetts and New Eng-

TABLE 4.1 Selected Data on Students Taking Scholastic Aptitude Tests [SATs] in 1984

	NATIONAL	MASSACHUSETTS	CALIFORNIA
Percentage of eligible seniors taking SATs	33%	63%	38%
Verbal score	426	429	421
Math Score	471	467	476
Parents' median income	$30,400	$28,800	$32,300
Estimated high school grade point average	3.04	2.86	3.09
Number of years of study of mathematics	3.65	3.80	3.51
Number of years of study of biological science	1.40	1.43	1.35
Number of years of study of physical science	1.86	1.91	1.43
% decline in verbal SAT score, 1971–72 to 1981–82	−5.8%	−6.2%	−8.4%
% decline in math SAT score, 1971–72 to 1981–82	−3.5%	−3.5%	−3.9%

SOURCE: The College Board, "College Bound Seniors, 1984," National, California, and Massachusetts.

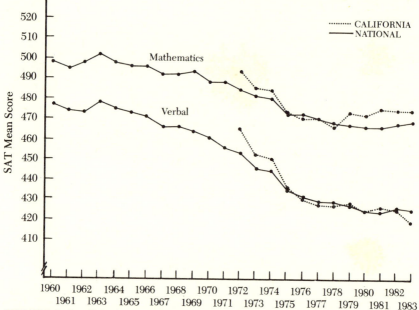

FIGURE 4.1   Scholastic Aptitude Test Scores for High School Seniors in the United States [1960–83] and California [1972–83]

SOURCE: California Assessment Program, California State Department of Education.

land students and, on some indicators, below that of all other regions of the country except the South.[4] A 1980 study of achievement was conducted for the Department of Defense from a national representative sample of 12,000 men and women ages eighteen to twenty-three. The results showed that those from New England scored highest on the Armed Forces Qualifying Test compared to young adults in eight other regions of the country while those from the Pacific Coast ranked near the bottom, above only the southern regions of the country. The most comprehensive evaluations of achievement are those of the National Assessment of Educational Progress (NAEP). Those studies have consistently shown students from the Northeast (which includes some Middle Atlantic states) to have the highest scores in mathematics and science followed by the Central, West, and Southeast states in that order.[5]

The California Assessment Program of the State Department of Education has a comprehensive annual statewide testing program where, unlike Massachusetts, students in all communities take the same test. The results from testing in recent years show that the average elementary-aged student in the state scores a little above or below national averages in reading, language, and mathematics. Eighth-grade pupils score considerably below the national average in reading but are close to the national average in writing and mathematics. Secondary school students perform below the national average in all three subject areas with average percentile scores in the 40s. Twelfth graders in the state scored close to the national average in 1969 when testing first began, but this was followed by a steep decline during the early 1970s, a decline that slowed in the mid-70s and flattened out at the end of the decade. Between 1981 and 1983 there was no significant improvement in test scores among twelfth graders in any subject area, but test results in 1984 showed slight gains in all fields.[6]

ACHIEVEMENT VARIATIONS WITHIN REGIONS

In both the Silicon Valley and Route 128 communities studied, there is considerable variation in student achievement. In both places students in more affluent school districts have higher achievement scores as is true throughout the nation. Generally speaking, students in Santa Clara County have scores on the statewide achievement tests that place them above state averages. The high school districts in the wealthiest areas in Santa Clara County—Palo Alto, Saratoga–Los Gatos, Los Altos–Mountain View, Fremont Union (which includes Sunnyvale and Cupertino)—have achievement scores that average from the 90th to the 99th percentiles on a statewide basis. The only secondary school district in Santa Clara County where the average student achievement is below the state average in most subject areas tested is the East Side Union District in San Jose, which serves a high proportion of poor and minority students. Students who attend schools in the more affluent sections of San Jose, which are included in the San Jose Unified School District and the Campbell Union High School District, have scores well above state averages. Achievement scores tend to be lower in schools in the more agricultural southern part of the county compared to the affluent districts to the north in the heart of Silicon Valley.

It is more difficult to compare academic achievement among school districts in the Boston area because school districts are free to choose their own testing program. Without comparable statewide test results, comparisons can only be made by looking at SAT scores. Since the proportion of students taking the SATs varies widely by community, direct and accurate comparisons cannot be made. However, the SAT results do indicate substantial achievement differences among school districts even when the test-taking proportions are taken into account. In more affluent cities and towns, such as Newton, Lexington, and Bedford, for example, students score on the average 45 to 58 points higher than state and national SAT averages in mathematics and 30 to 45 points higher in the verbal section of the test. In these communities a very high percentage of students, 76 to 82 percent, take the test so the scores are a fairly good estimate of student achievement in the district. In the working-class towns, such as Waltham, Billerica, and Woburn, where 52 to 54 percent of the students take the SATs, the average mathematics score is close to or a little above state and national averages but the average verbal score is 13 to 28 points below those norms. Thus, it is apparent that variations in achievement, which generally follow social class lines, are more pronounced within these two geographic areas than they are between them.

THE BOSTON PUBLIC SCHOOLS: A SPECIAL CASE

Special note needs to be made about the condition of the Boston public schools. Although few high technology manufacturing firms are located in the city, many are in nearby suburbs, and many service jobs in the city are technically oriented. Unlike the San Jose city schools, the Boston public school system serves largely poor and minority youth. About 70 percent of the 55,000 students enrolled in 1983–84 were minorities. Forty percent came from single-parent families, and 44 percent could be classified as low income (with parents who made less than $10,000 a year) according to 1980 data.[7] The high schools have a 50-percent dropout rate and student achievement is low:

. . . While students in the first grade start off at the national median, by the 10th and 11th grades, the aggregate system results show scores below the 40th percentile. . . . White students continue, as they always have, to score slightly above national medians in both reading and mathematics, as do Asian students. Black and Hispanic students, on the other hand, score very poorly in reading—by the third grade, the median Black reading scores are below the 40th percentile, and the median Hispanic scores are in the lowest quartile. The differential in mathematics scores is even greater.[8]

The system is characterized by contrasts. Teachers' salaries are relatively high, the pupil-teacher ratio is fairly low, and estimates on per pupil expenditures ranged as high as $6,400 per pupil in 1980–81. At the same time, there are serious shortages of certain kinds of staff, equipment, and supplies. Public support is low. A 1982 in-depth study by the *Boston Globe* highlighted some of the most glaring weaknesses:

Boston has the worst overall student attendance rate of any of the big-city schools nationally. . . . Almost one-quarter of the high school students are absent on a given day.

Nearly a third of the 10,000 high school students who took more than two academic courses during the second marking period this year [1982] failed at least half the subjects. With the exception of students at the city's three select examination high schools, 39 percent failed mathematics and 42 percent failed their science courses.

Vocational education programs are inadequate and under utilized.

More than one-third of the students enrolled in vocational education courses are failing their courses. More than half of the failures can be attributed to excessive absenteeism.

The number of sick days being taken by Boston teachers has been increasing during the past three years and last year was higher than those of 10 other cities surveyed.

Scores of classes are being taught by teachers who are not adequately trained for the courses they teach.

The number of guidance counsellors was reduced this year by 20 percent, even though their caseloads were already too heavy to allow adequate time with students.

Ten times the present budget is needed to repair and renovate the 132 school buildings.[9]

The school system was led between 1981 and 1985 by a reform-minded superintendent, Robert R. Spillane, but it still suffers from a long-term legacy of political corruption, patronage, and gross inefficiency. Despite some recent advances in organization, staffing, student and teacher attendance, curriculum, and test scores, enormous problems remain. The contrasts, then, between suburban school districts in the Boston area and the Boston public schools are stark and more pronounced than they are in Silicon Valley. Students in the city schools in San Jose are much less likely to be poor and disadvantaged and are served by an educational bureaucracy that is more efficient, better organized, and lacks the legacy of corruption that afflicts the Boston public schools. It is ironic that it was the San Jose Unified system and not the Boston schools that had to file for bankruptcy.

State and Local Differences in Mathematics and Science Course Taking

One of the explanations for the relatively good academic performance of New England students versus that of California students can be found by looking at the curricular patterns in the two areas. Students in high schools in New England and the rest of the Northeast are much more likely than students from other parts of the country to be enrolled in an academic curriculum. (Table 4.2) The fact that there was a significant migration of students into the less rigorous general curriculum during the 1970s was a source of concern to the National Commission on Excellence in Education. However, students in the northeastern part of the United States tended to stay with the more traditional academic course load. Only about a third of the high school seniors in other parts of the country were placed in an academic curriculum by 1980 but more than half (51 percent) of those in the Northeast (and 47 percent in the New England states) were in the academic track. Students in the western states, including California, were the most likely of all (45 percent) to be in the

TABLE 4.2　Percentage of 1980 High School Seniors in General, Academic, and Vocational Curriculum by Geographic Region

	NEW ENGLAND	NORTHEAST*	SOUTH	NORTH CENTRAL	WEST†	PACIFIC
Academic	47%	51%	33%	35%	34%	33%
General	25	24	39	40	45	47
Vocational	28	25	28	24	20	21

*Includes New England

†Includes Pacific.

SOURCE: *High School and Beyond*, National Center for Education Statistics, 1980.

general curriculum. According to the National Commission's report, "25 percent of the credits earned by general track high school students are in physical and health education, work experience outside the school, remedial English and mathematics, and personal service and development courses, such as training for adulthood and marriage."[10] What is particularly remarkable is that so many students who are planning to attend college opt for this curriculum. According to a study of the California State Department of Education, California high school students who are planning to attend community college are only slightly more likely than students planning to enter military service to take a more academic course load.[11]

Regional variations in enrollment in mathematics and science courses are also striking. High school students in New England are considerably more likely than those in other states to take three or four years of secondary mathematics and science while California students fall below national averages in mathematics and science course taking. (Table 4.3) According to the *High School and Beyond* survey of the National Center for Education Statistics (which does not have separate data on Massachusetts students), 65 percent of New England seniors graduating in 1982 took three or more years of high school mathematics compared to 46 percent of graduates nationally and 43 percent in California. And 46 percent of those in the

TABLE 4.3   Percentages of 1982 High School Graduates Taking Three Years of High School Science and Mathematics Courses, by Region*

	MATHEMATICS	SCIENCE
National	46%	30%
New England	65	46
California	43	17
Middle Atlantic	55	45
South Atlantic	49	31
East South Central	41	27
West South Central	44	25
East North Central	41	27
West North Central	46	30
Mountain	31	22
Pacific	42	18

*Courses taken during four years of high school.

Source: *High School and Beyond*, Transcript Study, National Center for Education Statistics, 1983.

class of 1982 from New England had three or more years of secondary science education compared with only 30 percent of their counterparts across the country. A mere 17 percent of California graduates had taken three years of science, a shocking figure.

California students average only 2.8 years of mathematics and 2.1 years of science during their four years of high school.[12] According to the *High School and Beyond* study of student transcripts, only 10 percent of 1982 Pacific Coast high school seniors had taken Physics 1 but 19 percent had done so in New England. (Table 4.4) Almost one-third of New England seniors had had a year of chemistry compared with only 18 percent of those in the Pacific region. (Table 4.4) Massachusetts students who take the Scholastic Aptitude Test have taken more years of mathematics and science than students in California and the nation, who take the test even though the test-taking population in Massachusetts is proportionally much bigger and therefore less selective.

It is the coursework of college-bound students in California that has been especially weak in the last decade. According to the Assessment Program of the California State Department of Education, one of the best research units in state education in the country, California college freshmen tend to have taken significantly fewer secondary courses than their collegiate counterparts across the na-

TABLE 4.4   Percentage of 1982 High School Graduates Taking Mathematics, Science, and Selected Courses

Course	National	Pacific*	New England
Algebra I	63%	68%	63%
Algebra II	31	27	53
Geometry	48	52	64
Trigonometry	7	6	12
Calculus	6	4	15
Physics I	11	10	19
Chemistry I	24	18	31
†Family Life or Sex Education	48	60	33
†Alcohol or Drug Abuse Education	39	51	31

*Reanalyses of this data show that the Pacific Coast sample results are almost identical with California results.

†1980 California, New England, and U.S. seniors.

Source: *High School and Beyond*, National Center for Education Statistics, 1983.

tion in all main academic subject areas.[13] (They were, however, much more likely to have carried a family life or sex education course and a class in alcohol or drug abuse education!) For example, in 1983, only 55 percent of college-bound seniors had four or more years of mathematics during four years of high school compared to 64 percent nationally. (One-third to one-half of the freshmen in California's public universities and colleges wind up in remedial mathematics classes each fall, many of whom have had a sufficient number of high school mathematics courses on paper but who failed to learn the material adequately.) More alarming, only 43 percent of these seniors took two or more years of physical science compared with 61 percent of college-bound seniors nationwide. A substantially smaller percentage of California students take the SAT science achievement tests than students nationally, a reflection of their lower course taking in these fields. For example, only 3,751 California seniors took the Chemistry Achievement test in 1984 compared to 4,419 in Massachusetts, a state with only about one-fourth as many high school students. A positive trend has occurred since 1981, however, with California college-bound seniors increasing their course taking in all core subject areas.[14]

There is some variation in course enrollments in mathematics and science in the Silicon Valley and Boston area schools. Milpitas High School in Santa Clara County requires four years of mathematics for high school graduation. Seven of the eleven high school districts in the county require only one year of science to graduate (one requires none) and six require only one year of mathematics, but in many of those schools, the college-bound students are already taking three or four years of mathematics. At Saratoga High School, for example, where most students are expecting to go to college, 62 percent of the students have taken three years of mathematics.

The statewide policies have recently been passed that will increase course enrollments in mathematics, science, and other academic subjects. First, the University of California has upgraded its entrance requirements, after lowering them in the late 1960s. Beginning with the high school graduating class of 1986, entering students must have had three years of mathematics instead of the two years previously required. The requirement of one laboratory course in science remains the same. These and other new guidelines, which require a heavier academic course load, have been criticized for being insufficiently stringent, but they clearly represent a stiffening of standards. The state college system began requiring two years of college-preparatory mathematics in the fall of 1984. Since such a large proportion of California students attend public higher education, these changes are having an important impact on students' course selections. Second, the California state legislature reinstituted state graduation requirements in the school reform bill of 1983. These requirements, which were abolished in 1969 and given over to local district discretion, include the mandate that all graduating students beginning with the class of 1987 have completed two years of mathematics, two years of science, three years each of English and social studies, and one of fine arts or a foreign language.

Massachusetts does not report precise data on the number of years of mathematics and science taken by its public school students, so accurate comparisons with course-taking patterns of California students are not possible. However, the SAT data reported above indicates that Massachusetts pupils probably average between three and four years of high school mathematics and three years of science during four years of high school. Graduation requirements, which typically require two years of mathematics and two years of science, are left up to local districts. The only state-mandated requirements for high school graduation are four years of physical education and/or health and one year of American history.

As in California, the public colleges and universities have recently moved to upgrade entrance requirements so that by 1987 entering freshmen will have to have successfully completed three years of mathematics, two years each of science (including a year of laboratory science) and social science, four years of English, and two years of foreign language. Since Massachusetts students are less likely to attend public colleges and universities than their California counterparts, and since most of them already exceed these requirements, the new rules will have less impact. In 1985 the Massachusetts legislature considered several measures that would have given the State Board of Education and State Department of Education much greater authority in setting curriculum requirements and approving school programs. However, the final legislation that was passed only enhanced the State Department's testing and data gathering function.

Enrollments in mathematics and science courses, usually prerequisites to professional technical careers, are relatively high in the eight high schools studied on Route 128. Sixty to 95 percent of the students in each school are enrolled in science courses at any one time. The figures for mathematics enrollments range from 80 to 99.8 percent; five schools have more than 90 percent of their students enrolled in mathematics courses. The popularity of mathematics and science is little related to the social and economic status of the community. In one of the blue-collar high schools, 90 percent of the students are enrolled in mathematics, 14 percent are carrying computer programming courses, and 95 percent are taking science subjects. Several schools had made a concerted effort to enroll non-college-bound students and females into science and mathematics courses. Five of the eight schools reported an increasing student enrollment in science courses during the past several years, and six schools reported similar growth in mathematics enrollments. Most of these increases are in the third-and fourth-year courses. No school was experiencing a drop in the percentage of its students in mathematics and science enrollments. These upward trends contrasted with those in Silicon Valley a year earlier (1980–81) where four out of eight school systems studied reported that mathematics and science enrollments were decreasing. However, follow-up interviews in several school systems in 1983 indicated that this trend was reversing.

These results show that Route 128 students go well beyond school district science and mathematics minimum requirements for graduation. Most of the systems require one or two years of high school mathematics and one year of science (two years for college preparatory students in several schools), but students take significantly more courses. Three or more years of mathematics courses and two to three courses in science are typical for the majority of students at these schools. One school that has no mathematics requirement for graduation has almost 100 percent of its students taking the subject. However, as student interest in science and mathematics grows, school budgets are being reduced and certain components of the curriculum are being dismantled, a process described in the preceding chapter.

Mathematics and Science: The National Scene

Although California and Massachusetts probably represent the extremes within the country on course enrollments in mathematics and science, there has been increasing public concern in the 1980s that the United States as a whole has inadequate school programs in those fields. The National Science Board's Commission on Precollege Education in Mathematics, Science and Technology has summarized evidence showing the decline and deficiencies in these pedagogical areas in the nation's schools. They point to various studies documenting the slide in science and mathematics achievement in the 1970s and the fact that remedial mathematics enrollments at colleges and universities jumped 72 percent between 1975 and 1980 even as total student enrollment at those institutions increased by only 7 percent. Other deficiencies were highlighted in the report: calculus is available to students at only one-third of U.S. high schools and two-thirds of high schools have physics courses taught by unqualified physics teachers; a small fraction of students (9 percent) in vocational education high schools take three years of science and only about one out of five take three or more years of mathematics; compared to twenty years ago, a smaller proportion of public high school students are enrolled in science courses; women and certain minority groups lag behind white males significantly in mathematics and science achievement; and curricula are seriously outdated and in need of revision in both method and content.[15]

International comparisons highlight deficiencies in mathematics and science education in the United States. A 1984 study by the National Center of Education Statistics shows that 1982 high school graduates in the U.S. had significantly less coursework in science and mathematics (an average of 2.2 years of science and 2.7 years of math during four years of secondary school) than students in West Germany, Japan, and the Soviet Union.[16] Preliminary reports from the Second International Study of Mathematics show the nation's twelfth graders tested in 1981–82 scoring below international averages in all areas of mathematics (although achievement of those enrolled in calculus classes was above the averages.). Eighth graders scored in the middle of achievement rankings among twenty-four countries in eighth-grade mathematics. But these American students ranked near the bottom when tested in measurement and geometry. The assessment showed that students in eighth-grade mathematics have only spotty exposure to topics in the curriculum, make only modest gains in knowledge over the year, and cover a good deal of material they already know. Only 10 to 15 percent take algebra, a low percentage compared to other countries. Eighth graders in this survey had lower scores in three out of five mathematics topics than U.S. eighth graders had in 1964.[17] However, a similar international assessment in science achievement (to be completed in 1986) found American students in 1983 knew more about biology and physical science than their counterparts did in 1970.[18]

There are some signs of positive change particularly among college-bound students. SAT scores have ended their erosion, which began in 1963. Students planning to attend college are taking more academic courses and more courses in mathematics and physical science in high school than ever before. The rise is especially noticeable for women, who have increased their course taking in mathematics at double the rate of men. In 1973, only 37 percent of the test-taking females anticipated completing four or more years of mathematics, but that figure rose to 59 percent in the 1984 group. Also, the latest mathematics survey completed in 1983 by the National Assessment of Educational Progress, which includes all students not just those planning to attend college, shows a halt in the decline of mathematics achievement.[19] It is likely that the increased time spent in mathematics courses is now paying off in stable or slightly increasing achievement scores.

Computers in Education

Among students and some teachers the current wave of enthusiasm about computer education contrasts sharply with the downbeat mood of educators in other curriculum areas. The growth in student use of computers has been swift indeed in the last few years and provides us with one school program area that is moving in a direction consonant with industrial trends. By the opening of fall classes in 1984, more than 85 percent of all public schools in the nation had one or more microcomputers for instructional use, up from 18 percent in 1981. The percentage varied by level of schooling with 93 percent of all high schools reporting microcomputer ownership (with an average of sixteen computers per school), compared with 82 percent of elementary schools (which average five computers per school).[20] The computers in high schools are used mostly for instruction in computer programming or as an introduction to computer courses, whereas elementary schools are more likely to use the computer for "drill and practice" instruction. [21]

Educators in Santa Clara County and in the Boston area, located in centers of the computer industry, have provided national leadership in organizing and promoting the use of microcomputers for educational purposes. This is a clear (and fairly rare) contemporary example of congruence between industrial and educational trends. The fact that Minnesota, one of the major centers of the mainframe computer industry, is also a leader in the adoption of computers for school use indicates that where concentrations of computer research and manufacturing exist, there is heightened activity and innovation in its applications in elementary and secondary schools. Unlike Minnesota, where direction and support for instructional uses of computers was centralized early at the state level, computer-using educators in Massachusetts and California (and most other states) had to develop their own programs at the local level. The adoption of computer curricula originated as a grass roots movement in these states. The state government in California began to put money directly into computer education near the end of Governor Jerry Brown's tenure in office in 1982, and now California ranks near the top among the states in its percentage of schools owning a computer. It is also now a national leader in the instruction of teachers in computer use. Brown was a vocal supporter of computer literacy in the schools and was able to push through a $26-million program, "Investment in People," which provided funds for the implementation of school computer programs and other technology initiatives. The legislation set up Teacher Education Centers (TECs) in regions across the state to serve as centers of leadership, training, and materials for public schools in the uses of computers. The California State Department of Education established a Computer Education Unit in 1983 to assist school districts in planning, implementation, and improvement of computer related curriculum and instruction. Additional state funds have been provided for educational computing since 1982.

The development of computer curricula in schools has been more uniform, planned, and extensive in Santa Clara County than it has been in the Boston area, although in both regions the major source of innovative, viable programs emerged at the outset in individual school districts such as Cupertino in Santa Clara County and Lexington on Route 128. The Santa Clara County Office of Education has had full-time staff people who have disseminated information, coordinated teacher in-service training, and developed a wide variety of other computer-related educational activities. School systems in the county have also had an additional option of paying a modest fee to join the county-run Computer Education Consortium, which provides them with services such as on-site teacher training, evaluations of software, and surveys of curricula. Most school districts in the county have joined. Critical state and national leadership has come from county educators, in part through the Computer-Using Educators group that was founded in the county and now has about 5,000 members in forty-nine states and twelve other countries. Teachers, administrators, and many parents have enthusiastically supported the adoption and expansion of computer curricula in the schools. Because the county education office and many of the local school systems had already been active in computer education, they were in a good position to take quick advantage of state money and support when it finally became available in 1982. This much-needed state support was readily utilized, particularly by such regions as Santa Clara County and nearby San Mateo County, which already had thriving programs.

Although Governor Michael Dukakis and the Massachusetts State Department of Education has been strongly committed to encouraging computer use in schools, significant state support has been slow in coming. Nor has there been a strong county system of government that would provide regional support within the state as is the case in California. Individual school districts have essentially been on their own in establishing and expanding computer curricula. As a result, Massachusetts schools have been somewhat less likely to be using computers for instruction than those in California. There are school systems that have comprehensive and well-thought-out computer programs, Lexington and Oxford for example, but there are others where very few students have access to such learning. The whole effort has been more decentralized and diffuse, even within the Route 128 area, than has been the case in Silicon Valley. There has been a proliferation of smaller groups, newsletters, and conferences in the Bay State, but no one organization or government agency has played a coordinating role. However, legislative proposals to provide state support and direction for school use of computers are likely to pass in the mid-1980s.

But like students in Santa Clara County, pupils in the Boston area have become extremely interested in computers. One secondary school system on Route 128 that had 231 students enrolled in computer electives in 1973 had 680 students taking such courses in 1982. Teachers at most of the eight Route 128 high schools studied have been quite willing, even eager, to retrain in order to teach these courses. Most of these teachers are in the mathematics departments, but interest comes from those in many other disciplines as well. Several districts have used creative methods to stimulate teacher retraining. For example, one school system had trained 375 school staff out of 700 (including some custodians and secretaries) in a thirty-hour course in BASIC through an Occupational Education grant administered by the state. Another system trained most of its mathematics faculty to teach computer courses through a "buddy system" where one teacher proficient in computers co-taught a course with a novice.

It is remarkable that in spite of the fact that there has been so little direct federal and state support for computer use in schools, there has been so much excitement and effort in this area of pedagogy. Parents and school adminstrators have purchased computers and software from a variety of sources. In Silicon Valley, for example, most of the school districts had to raise money from sources outside of the regular school budget, such as state School Improvement funds, Mentally Gifted Minor Programs, federal grants, or from parent association fund-raisers. Some school districts have purchased equipment from proceeds from the sale of recently closed school buildings, an option not legally available to school administrators in Massachusetts, where most computer purchases have been funded out of local school department budgets.

The fact that funding sources for the purchase of computers have been limited has meant that there has been a spotty quality to the rate of adoptions of computers for instructional purposes in both Silicon Valley and the Route 128 areas. Some towns have elementary schools (or a few elementary schools) with excellent programs, but there is little available at the junior high or high school level. In Santa Clara County, for example, the Cupertino elementary school district developed a nationally recognized computer curriculum, but for a time there wasn't any well-developed program their students could move on to in the Fremont Union High School District. It was estimated that half of the district's entering freshmen were computer literate since they were coming from the feeder elementary districts of Cupertino and Sunnyvale, which had also established computer curricula in the elementary grades. These students and their parents were successful in pressuring the high school district to develop and expand computer offerings by raising several hundred thousand dollars from the sale of school property in 1982. According to Bobbie Goodson, the computer resource teacher for the Cupertino district, the high schools have "really been prodded into this by the youngsters that were coming from our district expecting something."[22] That particular high school district now has one of the most well-developed computer education programs in the country, with computers being utilized in the teaching of a wide range of subjects.

California is moving in a direction that will make computer use more coordinated and equal among schools, but even there only 29 percent of California elementary schools had a microcomputer or terminal in 1982. And of those that did have a computer, 41 percent had only one and another 16 percent had only two. According to this statewide survey of computer use in schools in the spring of 1982, 78 percent of the sixth-grade students polled had never used a computer in school. Interestingly, this same study found that 18 percent of the students indicated that their family owned a computer and 45 percent owned a home video game.[23] Schools cannot satisfy student interest in computers. A 1982–83 survey by the Santa Clara County Office of Education on computer use in its middle schools found that teachers were facing "the problems of having too few computers available for the vast numbers of students wishing to use them. . . ."[24]

Similarly, while student interest in computers is "phenomenal," in the words of several administrators interviewed in the Route 128 area, funding to accomodate student interest is inadequate. Officials interviewed at all of the eight high schools said that financial constraints had prevented them from acquiring the computer hardware necessary to meet student demand. Like students in California and elsewhere in the country, these pupils do not have very much exposure to computers even though interest is very strong. According to the Johns Hopkins University survey in 1982, students in the United States have the following experience with microcomputers in their schools:

The typical microcomputer-owning secondary school has approximately five microcomputers, each in use for 13 hours per week, or a total of 65 hours of use. About 80 students (in a student body of 700) use the equip-

ment in an average week—a little more than 45 minutes per user. . . .
The typical microcomputer-owning elementary school has two microcomputers, each used for about 11 hours per week, or a total of 22 hours of
use per week by students under the direction of a teacher or other staff
member. About 62 students (in a student body of 400) share these 22 hours
of use, which is equivalent to about 29 minutes per user per week. [25]

There is no question that if a stable and increased source of funding for computers were forthcoming from the state or federal
governments, schools would adopt computer programs at a much
faster rate than is currently the case. Teacher resistance to such curricular changes is a real phenomenon (since a real commitment on
their part to learn the new skills is required) but has been greatly exaggerated. In the Cupertino and Evergreen school districts, approximately 80 percent of the teachers have had at least one computer
class. According to William J. Wagner, a national figure in computer education and coordinator of computer education in the Santa
Clara County Office of Education, teachers may be a little reluctant
at first but within a year or two large numbers of them are eager to
familiarize themselves with computer instruction:

There has been an explosion of interest among teachers, some of whom are
in their 60s. Some of them are coming out of the woodwork whom I never
dreamed would be involved. Eight hundred teachers, most of them from
this county, took computer classes this spring and another 1,300 have
signed up for summer courses. It just took teachers a little while to catch
on. Teachers *are* making a big difference in this revolution.

The same phenomenon of teacher involvement was observed in the
Route 128 schools.

The computer revolution has caught hold of educators without
its being forced on them by school administrators. Within the
general educational context of austerity and despair, the use of computers in education is providing a source of innovation and excitement among school staff. There are, of course, many unanswered
questions about the form computer education should take,
especially at the elementary school level, and its utility to students. It
is still too early to assess the long-term learning outcomes of various
styles of computer use in the classroom. It is clear, however, that
many kinds of employers want high school graduates to have some
acquaintance with the use of computers, which does not necessarily
include programming skills. And college-bound students are at an

advantage in their scholastic studies if they have some computer expertise.

Teachers: National Trends

The most serious problem facing American education today and in the future is the condition of the teaching force in elementary and secondary schools. The issue of the quality and quantity of new recruits to teaching now occupies center stage in the debates about educational reform. The creation of incentives as well as sanctions to improve teacher performance has become a major political issue as well. Several major problems beset the teaching profession in the mid-1980s: the shortage of newly trained teachers; the decline in academic qualifications of new teachers and teacher-trainees; and the widespread demoralization of veterans on the job. The extent of these difficulties varies from region to region of the country, but all states are facing these issues to one degree or another. As states in the mid-1980s implement requirements for more rigorous academic curricula, they are finding an inadequate number of trained people available to teach those courses. The extent of the teacher supply issue differs by subject area with more serious shortages occurring in the fields of mathematics and science and vocational education.

The teacher work force began to decline only in the late 1970s, several years after student enrollment drops had begun. School districts initially adapted to declining enrollment by reducing class size and utilizing staff to service mandated special education and other specialized programs. Later in the decade, when budgets shrank and the number of students continued its drop, school districts reduced the size of their teacher force; nationwide, the number of teachers declined from 2.49 million in 1977 to 2.38 million in 1984. The demand for elementary school teachers began to rise in 1985 as the number of students started to increase again, and by 1990 there may be 20 percent more elementary school teachers than there were in 1980. Since enrollment reductions will still be working their way through secondary schools throughout the 1980s, the number of secondary teachers is not expected to increase during that decade.[26] Shortages of teachers will be a particular problem in the southern and western states where the population is growing. California educators, for example, are predicting massive teacher shortages by the early 1990s as a result of retirements, resignations, and enrollment increases.

There has been a dramatic decline in recent years in the number of newly trained teachers. Between 1971 and 1980, the number of new teacher graduates as a proportion of all bachelor's degrees dropped from 37 percent to 17 percent. Bachelor's degrees in education declined by 39 percent between 1972–73 and 1980–81 with the percentage reduction being greater for females than males. The drops were most pronounced in the fields of general elementary education, art, home economics, business, and mathematics.[27] Reductions in prospective education majors reached an all-time low with the college freshman class of 1982: only 4.7 percent of the freshmen polled nationwide claimed they wanted to become teachers, down from 22 percent in 1966. The drop was most striking in secondary education with only 2 percent of the freshmen indicating aspirations to that field. (Figure 4.2) Data from 1984 classes indicate, however, that the drop in education majors may have stabilized. A slight increase in education was registered by college-bound students taking the SATs in 1984, ending an eight-year decline. Some colleges reported modest growth in teaching-related enrollments in 1984–85.

The slide in academic achievement affecting all groups of students appears to have been more precipitous among those preparing for a career in education. For example, college seniors taking the SATs who selected education as their intended field of study in college had a steeper absolute and proportional drop in their SAT mathematics and verbal scores than did students overall. Between 1972 and 1982, SAT verbal scores for prospective education majors fell by 24 points compared to a 19-point drop among all other students while their mathematics scores fell even more, 30 points, contrasted with a 14-point decline among the rest of the test takers. Average academic achievement of students preparing for teaching careers has always ranked fairly low compared to that of students selecting other fields, but as a recent National Institute of Education report put it, "a mass of evidence converges to show that academic ability of education majors is both low and declining."[28] The gap that has always existed between the academic achievement of teacher-trainees and students in other fields has grown even wider in recent years with teaching attracting fewer and fewer of the high-ability students. Further, other research has shown that those academically able students who do train to be teachers are more likely

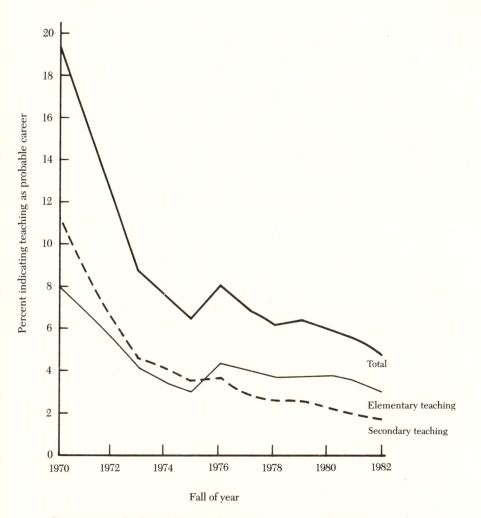

The proportion of college freshmen indicating elementary/secondary teaching as their
probable career declined throughout the 1970s, dropping to under 5 percent in 1982.

FIGURE 4.2   College Freshmen Indicating Teaching as Probable Career

SOURCE: American Council on Education, Cooperative Institutional Research
Program, *The American Freshman: National Norms,* various years.

than low-ability pupils to choose not to teach in the first place or, if
they do, to defect from the occupation by the time they are age
thirty.[29]

The issue of teacher supply has become especially critical in the
fields of mathematics and science. There has been a chronic shortage
of teachers in these areas for decades, but the supply of such

instructors has become even more reduced in the last decade as better-paying jobs in industry have opened up for people with skills in those disciplines and as the attractiveness of teaching has declined. Nationwide, there was a 65 percent decline in science education graduates and a 77 percent decline in mathematics education graduates between 1971 and 1980. Most states are reporting a shortage of mathematics and science teachers with demands for physics teachers being especially acute. Only half of the student teachers in the 1970s in the fields of mathematics and science education actually entered teaching, and one-quarter of those currently teaching in those fields claim they plan to leave the profession in the near future. Approximately five times more mathematics and science teachers resigned from teaching jobs in 1981 in order to take jobs outside of teaching than left because of retirements. As a result of the shortage of qualified personnel, half of the newly hired teachers in secondary mathematics and science in 1981 were not certified to teach in those fields. And up to 30 percent of all science teachers currently in the classroom appear to have been assigned to teach courses for which they are inadequately trained.[30] Ironically, the increased secondary mathematics and science course load now being implemented in most states is making the problem worse. Senior teachers in other fields who are losing students because of declining enrollment or new curricular mandates are being pushed into science and mathematics classes.

Another problem is the demoralization of a large percentage of the current teaching staff. The profession is low paid—the average salary was only $23,546 a year in 1984–85 after an average of 14 to 17 years of experience. Graduates in business administration start at salaries about $3,500 higher and the gap widens significantly between the two fields as entrants gain increased experience.[31] The job security that used to characterize the field has been lost as thousands of veteran teachers face layoffs each year. Those remaining on the job have few new colleagues to work with: in 1971, 32 percent of all teachers had four or fewer years of teaching experience but that proportion had declined to 14 percent by 1981.[32] By 1984, the average age of teachers had climbed to about forty–one nationwide (several years higher for science and mathematics teachers) as younger instructors lost their jobs and new hiring slowed. Although the turnover rate among teachers has declined from 8 percent in 1969 to approximately 6 percent in the 1970s and 1980s, this is not a reflection of greater satisfaction with the job.[33] A National Education Association survey in 1961 found that 50 percent of teachers would definitely choose that occupation over again if given the choice. However, by 1983, only 24 percent, when asked, said they would choose that same field.[34]

The Teaching Profession in California and Massachusetts

California has been affected more seriously than Massachusetts by teacher shortages and the assignment of unqualified teachers to cover certain subject areas. These problems are significantly less severe in the Northeast than they are in the rest of the country, particularly in the areas of secondary mathematics and science. The demand for new teachers is lower and the percentage of newly employed but unqualified science and mathematics teachers is sharply lower—9 percent—than the national average of 50 percent. But the proportion of new but unqualified hires in mathematics and science in the Pacific Coast states is an astounding 84 percent.[35] A survey by the California Science Teachers Association in 1982 found that 50 percent of the state's high school math classes and 72 percent of the junior high school classes were taught by people who did not major in mathematics. Two-fifths of these junior high teachers did not even minor in mathematics.[36] In all of California's teacher training programs in 1982, there were only 184 students preparing to teach science and 110 preparing to teach mathematics, leaving a shortfall of 1,600 to 1,700 qualified teacher applicants for the 2,000 teaching vacancies in those fields statewide.[37]

In Massachusetts, according to a 1983 report of the State Department of Education, there was no technical shortage of mathematics and science teachers statewide because "a shortage doesn't exist without a demand, and there is apparently no current *general* demand for math and science teachers" although a shortage was projected for 1985. In 1981, after the passage of Proposition 2½, 253 mathematics and science teachers were laid off from their jobs, and terminations are still continuing in some districts. Meanwhile, enrollments of teacher-trainees in mathematics and science have plummeted, and, overall, the number of bachelor's degrees conferred in education dropped by almost 40 percent between 1971 and 1981 in the state. (Table 4.5) A total of eighty-five students trained

TABLE 4.5   Number of Graduates in Secondary
Mathematics and Science Education in Massachusetts
Colleges and Universities, 1983, 1984

SUBJECT	1983 GRADUATES	1984 GRADUATES	TOTAL
Math	50	60	110
Math & science	7	10	17
Chemistry	2	4	6
Physics	2	2	4
Biology	24	22	46
Earth science	0	1	1
Total	85	99	184

SOURCE: Massachusetts State Department of Education.

to teach in mathematics and science at the secondary level (which requires that they major in mathematics or in a science field) graduated from Massachusetts public and private colleges and universities in 1983 and approximately 100 did so in 1984. Among the 1983 graduate group, there were only two graduates each in the fields of physics and chemistry. The state report cautions that some of these graduates will not choose to teach, some will leave the state (Massachusetts has been an exporter of teachers for years), and very few will enter the state from other parts of the country.[38] A 1982 survey of high schools by the Massachusetts Science Teachers Association and the State Department of Education revealed that more than half of the responding schools are having difficulty in hiring qualified candidates in science education, particularly in the fields of physics and chemistry. And the survey indicated that many science courses across the state are being taught by unqualified teachers. A 1983 statewide study concluded that more than one-fifth of the junior and senior high schools had one or more inadequately trained science teachers.[39]

The situation in neighboring New Hampshire, home now to many high technology firms, is much more severe. The average teacher's salary there in 1984–85 was a paltry $18,577 (placing it forty-eighth among the states), which has led to a drastic shortage of qualified teachers. Fewer than ten certified teachers in mathematics and science are graduating from New Hampshire colleges and universities each year. More than two-thirds of those teaching physics and almost half of those teaching chemistry are not certified in those fields.[40] The state now requires high school students to take a computer literacy course (as do about half the states as of 1985) but schools are struggling to find qualified teachers to instruct the courses. Maine also is experiencing teacher shortages as a result of its low pay scale.

It appears also that shortages of teachers for all subjects may appear in Massachusetts in the near future: only 3.8 percent of the state's 1984 high school seniors taking the SAT's (two-thirds of all seniors) selected education as their intended area of study, compared to 4.6 percent nationally, and a mere sixty-four students selected secondary education. Since the overall demand for elementary and secondary teachers is expected to decrease slightly between 1980 and 1990, the decline in teacher-trainees might seem to be a positive trend. However, replacement needs in the state created by retirements and resignations, estimated to be 1,600 positions annually in the elementary fields and 1,100 positions each year in secondary disciplines, will not be met if present career trends among college students continue.[41]

The quality of new recruits to teaching has also become a major issue in both Massachusetts and California. The latter state instituted a teacher competency examination in 1982 for veteran teachers seeking new certifications and newly graduated college students seeking certification for the first time. During the first six months the test was given to full-time veteran teachers, 27 percent failed. Among college students training in education (who in California must already have completed four years of college), 32 percent failed during the first year of the test's use. (Six percent more passed on a repeat test.) About half of the Hispanic candidates in this group and more than two-thirds of black test takers were unable to pass the test. In 1985, San Jose State University began requiring applicants to its school of education to pass the state test prior to admission. Students entering the teaching program must also have at least a B-minus undergraduate grade point average. Beginning in 1986, students admitted into the education program in the California State University system must have ranked in the upper half of students in their discipline. There no comparable test in Massachusetts, but the SAT scores of prospective education majors in 1984 were very low—391 in the verbal section of the test and 414 in mathematics.

SILICON VALLEY TEACHERS

The school systems in Santa Clara County and the Boston area are no exceptions to national state trends. Fiscal austerity measures and declining enrollment have caused extensive layoffs of younger teachers, including mathematics and science instructors, in Silicon Valley. Approximately 1,100 professional school staff (teachers, counselors, librarians, nurses) received layoff notices in Santa Clara County in 1982: Only thirty of the teachers were hired back on a permanent basis the following year. Several of the area's school districts have laid off teachers hired as long ago as the late 1960s. The average age of teachers in the Cupertino Union School District, which had significant teacher layoffs in 1984, is now forty-eight. Terminations were continuing in many districts in the county through 1985.[42] A few districts, such as San Jose's East Side Union High School District, which have growing or stabilized enrollments are hiring teachers but these systems are the exception. Programs that rejuvenate veteran teachers, such as sabbaticals and in-service training courses, have been sharply reduced.

School officials who do hire teachers on a permanent or temporary basis report difficulty in finding qualified people in the fields of mathematics and science but have generally avoided hiring people in these areas on an emergency credential basis—which means they are not certified to teach that subject. However, in all the eight high school districts studied in the Valley, veteran teachers who were not truly qualified to teach mathematics and science courses have been assigned to instruct them. Emergency credentialing can be bypassed since teachers certified prior to 1973 received a general secondary teaching credential, which enabled them to teach in disciplines other than their college major. This credential allowed the assignment of many unqualified teachers to mathematics and science courses.

During 1984, thirty-three Santa Clara County veteran teachers enrolled in a summer program at the University of Santa Clara (funded in part by Chevron Corporation) to retrain to become mathematics and science teachers. Across California, several hundred teachers a year are involved in such retraining efforts, which range in length from a few weeks to fifteen months. This assignment of "retreaded" tenured teachers into mathematics and science classes is a stopgap effort to cope with teacher shortages in those fields. But the utility of programs that provide only brief reeducation experiences for instructors is questionable.

Local colleges and universities are producing only a handful of newly trained teachers in secondary mathematics and science: in 1982, Stanford, the Universities of California at Berkeley and Santa Cruz, San Francisco State, and San Jose State graduated only fifteen people in secondary mathematics and thirty-two in secondary science.[43] Six of the eight high school districts studied in 1981 were losing mathematics and science teachers to better-paying jobs in industry, an exodus that continues in some districts. This is true even though teachers' salaries in school districts in Santa Clara County are relatively high by state and national standards: In 1982–83, for example, teachers' average salary in San Jose Unified was $28,114 (rolled back to $27,545 after bankruptcy was declared) and averaged from approximately $30,000 to almost $33,000 in Palo Alto Unified, Fremont Union, Campbell Union, and Santa Clara Unified school districts. Overall, the average pay for Santa Clara County teachers in 1982–83 was $27,318.[44] By contrast, teachers' salaries in even the more affluent Route 128 suburbs of Boston averaged only between $23,000 and $25,000 during that same year. Beginning teachers in Silicon Valley can expect to earn at least $3,000 more a year than their counterparts in the Route 128 area.

ROUTE 128 TEACHERS

A more extensive study of the state of the mathematics and science teaching staff in eight Route 128 high schools in seven communities in the Boston area was conducted as part of this study in 1982. These schools reside in the center of the Route 128 electronics and computer complexes. Two of the communities are upper middle class, another two are middle class, while the other three are largely blue collar in their socioeconomic makeup. At least one high-ranking administrator in each of the systems as well as the chairs of the sixteen secondary mathematics and science departments were interviewed. Questionnaires were sent to all 241 mathematics and science teachers in the eight schools, 66 percent of whom (158 people) responded.

Although Massachusetts has much less of a shortage of mathematics and science teachers than California and other parts of the country, it was apparent from the interviews and questionnaire data that the supply of qualified mathematics and science teachers is also becoming a serious problem in the Boston area. All the department chairs interviewed stated that the pool of candidates was very small and five of the eight schools reported difficulty in finding qualified teachers in those subjects, especially physics. "Five years ago we had over two feet of job applicants' folders and now we have one inch," reported one mathematics department chair. One school system had experienced an 88 percent drop in applicants in all teaching fields in the last few years. "We are going to have a terrible shortage of teachers," warned the assistant superintendent. Many administrators expressed concern about the quality of new recruits. "We're getting the crumbs," commented one science chair. Another administrator admitted that "today we'll take any warm body" to teach mathematics and science.

Most of the Route 128 schools studied do not have student teachers in science or mathematics, or other subjects for that matter, and many schools no longer have Future Teachers clubs. Able students are no longer encouraged by their teachers, parents, or friends to choose education as a career. The number of students graduating in science and mathematics in local colleges and universities is miniscule. Nearby Lowell University has abolished its undergraduate secondary education department. Half of the new certifications in mathematics in 1981 in Massachusetts were veteran teachers "adding on" an additional certification, by taking six additional courses, in order to save their jobs. These "add-ons" are not truly qualified to teach more than an introductory course.

Unlike Silicon Valley, officials at the Route 128 schools studied had not yet resorted to assigning unqualified teachers already on the staff to teach these courses. However, misassignment did appear to be a growing problem in junior high schools where senior elementary teachers with a K-8 certification were "bumping" less experienced junior high math and science specialists. This was the case in four of the seven school systems studied. Several administrators believed that these senior elementary teachers were clearly less qualified than the specialists they had replaced. Many expressed the fear that misassignment would become a problem at the secondary level in the future as it already has in the city of Boston and other districts in the state. Boston is experiencing a severe shortage of mathematics teachers. The problem there has been compounded by fairly widespread replacement of qualified mathematics teachers by more senior teachers transferred from other fields.

Staff attrition is a serious problem in two of the eight schools. One of the two had lost five math teachers in the last three years to jobs in the higher-paying high technology sector. Another three high schools were losing approximately one a year to such jobs. Three schools, those in the most affluent communities least affected by budget cuts, claimed that such losses were not a problem, but the chairs predicted that it might become so in the near future. Their fears were borne out in the questionnaire survey of teachers, which revealed widespread demoralization among the mathematics and science staffs. The teachers, over two-thirds of whom are male, averaged sixteen years of experience in 1982, the same as the national average for secondary mathematics and science teachers. By national standards, they are an extremely highly educated group: 81 percent, compared to approximately 50 percent of the secondary mathematics and science teachers nationally, have an advanced post-graduate degree, and many have further coursework beyond that.[45] All of the science chairs and two-thirds of the mathematics chairs had participated in the National Science Foundation programs of study available in the 1960s and early 1970s.

One of the sources of teacher malaise is apparent from the fact that only 7 out of 158 teachers in the 1982 survey were under the age of thirty. Only two of the respondents had four or fewer years of teaching experience. One science department did not have a single teacher under the age of forty. This age distribution is the source of much concern among those surveyed. According to the study results, only 41 percent of the mathematics and science teachers in the Route 128 high schools were planning a permanent career in teaching or school administration. The rest were either expecting to be laid off (12 percent) or were seriously considering leaving teaching (47 percent). The proportion expressing such feelings varied by school: in the system with the largest budget cut, every math teacher who responded to the survey wanted to leave teaching; in another system, two-thirds of the science teachers wanted to remain in the profession. Of course, teachers have always had high rates of attrition, but much of that was traditionally due to females in their twenties leaving teaching when they had children and males moving up to administrative positons. But these Route 128 schools have senior staffs who would normally not be expected to have such a high projected turnover. Thus, the fact that almost three-fifths of them are seriously considering a career change is cause for some alarm.

The percentage of mathematics teachers wanting to leave was the same as that of science teachers. Of physics teachers, however, six out of nineteen in the sample said they wanted to continue their teaching careers, a lower fraction than that of teachers in other mathematics and science fields. Ironically, at a time when there is such a shortage of physics teachers, four of these teachers believed they were going to be laid off. (Nine physics teachers were laid off in 1981 statewide according to the State Department of Education.) Several chairs commented that when mathematics and science teachers are "riffed," few try to get another teaching job. "It's the elementary teachers who hang on," said one school administrator. Teachers who were younger and less experienced were more likely than others to be thinking about a career change. Of the seven teachers under thirty, only two believed they would stay in teaching. Of those who were considering a career change, 60 percent said they would seek a job with a high technology company or a job that in some way utilized their mathematical or scientific expertise (often in computers). Their reasons for wanting to leave were twofold: declining job satisfaction and inadequate salary compensation. Typical comments written on the questionnaire included:

"I'm tired of the long hours and low pay. Nobody except teachers care about the education of children."
"Teaching is now a dead-end occupation."
"I'm tired of teaching and the future looks bleak."
"The teaching environment becomes less pleasant each year."
"I love teaching but it's wearing me out. There are too many alienated kids to face every day."
"Working conditions are deteriorating. I want to leave Massachusetts and its anti-education attitude."
"Teaching is a luxury I can no longer afford."
"There is no room for growth, no reward to excellence in teaching. Salaries do not keep pace with inflation."

The most serious problem, then, in high school mathematics and science instruction in these Route 128 high schools is the dissolution of a highly qualified cadre of teachers with no new cohort of trained people to take its place. Statewide, more than two-fifths of secondary schools report that loss of experienced teachers in science is a major problem.[46] With teaching salaries beginning between $11,000 and $17,000, students with skills in these fields can often find alternative employment at almost double that figure. Job security, once a strong point of teaching, no longer exists. Sabbaticals and national curriculum development programs, which these teachers found rewarding in earlier years, have all but disappeared. Moreover, the veteran teachers surveyed felt that the quality of students' work was not improving—indeed, two-fifths felt that student academic performance had slipped in the last five years. With equipment cuts, larger classes, reductions in professional development opportunities, it is not surprising that teachers are demoralized. The shortage of qualified teachers in all fields will become more and more apparent as a large number of teachers hired during the boom years of the 1950s retire, when enrollments begin to rise at the elementary level, and as the full impact of the decline of newly trained teachers and the exodus of veteran teachers is felt.

INDUSTRIAL ARTS AND VOCATIONAL EDUCATION TEACHERS

There is a shortage of new recruits to vocational and industrial arts teaching, particularly in high technology areas like computers and electronics, and there is a simultaneous loss of experienced teachers in these fields to industrial jobs. All of the seven school systems studied in the Route 128 area and two regional vocational high schools on Route 128 are having difficulty hiring instructors for electronics and computer courses. In one high school, all three electronics instructors had left over a three-year period for jobs in private industry where they made substantially higher salaries. Vocational educators are paid $13,000 to $27,000, a scale that cannot compete with industrial compensation. Although salary scales are higher in Silicon Valley (due in part to passage of a statewide minimum salary of $18,000), the same problem exists in their vocational and occupational programs where qualified teachers are scarce and turnover is high.

The problems of finding instructors and keeping veteran teachers up to date in changing technologies are occurring at a time when vocational students are gravitating to technology-oriented programs. Computer-related courses are popular at all schools. A study in Massachusetts found that secondary enrollments in vocational programs have been shifting in the direction of high technology courses such as computer-related business and office programs and electronics assembly and technician training. Between 1972 and 1979, there was more than a 110 percent increase in enrollments in high technology–related occupations although total vocational enrollments rose only 50 percent.[47] A more recent study by economist Patricia Flynn found that in the area of Lowell, Massachusetts (in the Route 128–495 belt), there has been a significant enrollment upturn in these technically oriented fields.[48]

In Silicon Valley there are some strong industrial arts and vocational courses relevant to careers in electronics firms, such as the nationally recognized electricity-electronics curriculum of San Jose's East Side Union High School District and a variety of programs run by the three Regional Occupational programs in Santa Clara County. Vocational students in the southern part of the county are now flocking to electronics courses as high technology firms expand into that agricultural region. But attractive industry salaries continue to depress the supply of competent technology-oriented faculty in this and other fields.

Efforts and Attitudes Toward Change

A movement to institute higher academic standards in schools has swept the country in the 1980s. In many states the push for reform has been coupled with a commitment to increase funding for education, although this commitment remains uncertain and fragile. Federal funding for mathematics, science, and technical education has resumed on a limited basis after years of neglect. The National Science Foundation, whose pre-college science education programs were decimated in the first two years of the Reagan administration, is now running teacher institutes that train both outstanding veteran mathematics and science teachers as well as teachers whose skills in those fields were deficient to begin with. Other kinds of in-service training programs for mathematics and science teachers have also resumed with federal funding made available through the 1984 Math/Science bill (Education for Economic Security Act). Some federal funds from the Math/Science bill are going to demonstration and exemplary programs in mathematics and science instruction at the local public school level and for traineeships for college undergraduates planning careers as mathematics and science teachers at the secondary level. However, the sums allocated during 1985 for all activities under the Math/Science bill amounted to less than $2 per child per year.[49] The amount of paperwork required to apply for these small sums has discouraged many school districts from applying for the funds.

Changes implemented in California are a good example of the range of policies and programs being established to reform and refund public education at the state level. There has been more ferment in California because the problems there had reached a crisis stage and because its high per capita income (fourth highest in the nation) made its educational decline more disgraceful than academic deficiencies in poorer states. Efforts for change in California came first from leaders in the educational and academic communities and eventually came to include the business establishment as well, acting through the California Business Roundtable. The failures of California schools received a good deal of publicity during the campaign in the fall of 1982 for superintendent of public instruction. The winner, Bill Honig, who had the backing of the business community, had stressed the importance of upgrading standards and linking these reforms to additional financial support for schools. Finally, after extensive media attention to the inadequacies of the state's schools, California voters expressed the view that they were willing to pay higher taxes for school improvements.

The turnaround in the attitudes of California citizens toward school funding occurred during 1982 and 1983, as several school districts teetered on the verge of bankruptcy and as national reports on the condition of American education became front-page news. Although Proposition 13 had slashed school spending, it remained popular with the state's voters. By spring of 1983, however, a Field poll of residents found that they were split in their assessment of its impact with 43 percent saying that the tax reduction measure had left them neither better nor worse off, 29 percent claiming they were better off and 23 percent reporting they were worse off. Parents of children in public schools or in state colleges and universities were most likely to report that Proposition 13 had had an adverse effect.[50] In a survey of residents in the city of San Jose by the *Washington*

Post, in September 1983, 53 percent of the respondents said that Proposition 13 had worked out unfavorably for the people of San Jose compared with 34 percent who said its effects were favorable.[51]

Polls released in June and July of 1983 showed that California citizens held their public schools in very low regard (only 8 percent said their local public schools were doing an excellent or a very good job compared to 29 percent in 1967) and were willing by an over-whelming margin (64 percent to 31 percent) to see their tax bills rise to pay for better public schools. In 1982, a similar poll found that only 41 percent of California adults believed that schools could be improved by spending more money. But by the spring of 1983, that proportion had jumped to a remarkable 66 percent, "an almost complete reversal" according to pollster Mervin Field.[52] Another poll found that residents of Santa Clara County were even more likely than residents in the rest of the state to support more spending for public schools (77 percent compared to 71 percent). Large majorities in the county (75 percent) and the state (67 percent) wanted to lengthen the school year and to toughen the curriculum (74 percent in the county and 68 percent in the state).[53]

The California legislature took the first steps toward upgrading school standards and improving the schools' funding condition when it passed the Hughes-Hart Educational Reform and Finance Act of 1983 (SB813), a comprehensive educational package of eighty reforms. Obvious public support for the proposals prevented Governor George Deukmejian from vetoing some of the financial portions of the package. The Reform Bill, along with successive state school budgets, provided for increases in school aid to purchase textbooks, raise starting teacher salaries over a three-year period to $18,000, restore some of the cuts made in counseling staffs, and pay "mentor teachers" (a few per district) an additonal $4,000 a year to take on extra tasks. The bill also eliminated the lifetime credential for beginning teachers, allowed school boards to make exceptions to seniority rules in laying off, reassigning or rehiring teachers, required competency tests for all teachers assigned to positions for which they do not hold credentials or college majors, and made it easier for districts to dismiss incompetent teachers. State high school graduation requirements were also reimposed. The package also lengthened the school day and year beginning in the fall of 1984 and reinstituted summer school classes in mathematics and science in that same year.

School funding is also being supplemented by revenues being generated by a statewide lottery approved by voters in 1984.

Entrance requirements to the state's public four-year colleges and universities have also been stiffened, requiring that students take a heavier academic load, and many local districts have started requiring a C average for students who wish to participate in sports and other extracurricular activities. Almost all Santa Clara County high school districts have toughened their academic eligibility standards for such participation. Further, state superintendent Bill Honig began implementing a comprehensive accountability program for all the state's public schools in 1985 that assesses schools' effectiveness on a variety of indicators. A set of detailed curricular goals and guidelines for high school classes statewide has also been introduced by the State Department of Education.

These educational changes represented a real reversal for California education. Yet they should be regarded as only the beginning of what must be a sustained wave of change if long-term upgrading of the public schools is to take place. The implementation of almost all of the reforms that have been passed costs money. One estimate is that California will have to spend $2.75 billion annually to bring the state's schools up to the national average in the amount of time spent in the classroom and in class size. And even when the reforms in instructional minutes are fully implemented, California students will still fall a little short of current national averages in class time. Class sizes remain large by national standards. And it will cost a good deal of money to provide the additional teachers and facilities needed for the academic courses being required for graduation in 1987. The task is daunting as only 40 percent of 1982–83 California seniors fulfilled the newly mandated graduation requirements in science and only 61 percent did so in mathematics.[54] Although the 1984 through 1986 state school budgets provided significant additional funds for public education, in reality the budget increases for individual school districts were not enough to remedy the problems the schools face. Adequate funding will have to exist for a period of years before the schools are really back on their feet. Whether the political commitment for that exists remains an unanswered question.

There is a similar movement for educational reform in Massachusetts, although it is being pursued with less urgency than is the case in California. The public colleges and universities have toughened entrance requirements, which take full effect in 1987, and have stiffened teacher certification standards. An educational reform bill passed the state legislature in 1985. (A more comprehensive and expensive package of changes passed the State House of Representatives the previous year, but then the bill foundered in the State Senate.) The reforms, funded by an increased appropriation of $211 million over two years, include augmented state aid to poorer school districts, systematic and uniform assessment of student skills, more stringent teacher evaluation methods, small increases in teachers' salaries, development of pilot pre-school programs, and the establishment of local school building councils authorized to spend money on school improvements at the rate of $10 per pupil. Approximately one-fourth of local school districts have already raised high school graduation requirements to take effect in 1987. Several of the state's colleges and universities have begun or have reinvigorated programs to train secondary mathematics and science teachers.

There are also proposals to expand and stabilize the state's share of public school expenditures, proposals that are supported by Governor Michael Dukakis. However, Massachusetts citizens, with the exception of those in poor and older cities and towns that have been hit hardest by the effects of Proposition 2½, do not feel that the schools are in a crisis condition as they are in California and some other states. And legislative leaders are split on the amounts of new monies needed and on the necessity for a tax increase to raise the additional funds needed to implement school reforms. Further, Governor Dukakis opposes a tax increase to cover major school changes. Thus, even though recent polls show strong public approval for school improvements of various kinds, it may be difficult to rally political support for additional revenues and substantially increased expenditures for schools.

California and Massachusetts as well as many other states appear to be reorienting their educational systems in a direction that will better prepare students for jobs in a more technologically oriented society. However, this reorientation is a long-term process and it is premature to predict the outcome. The raises in teachers' salaries being passed or proposed in some states, for example, may be too small to attract high caliber candidates to the profession. A 1984 study predicts that the wide gap between the salaries of mathematics and science teachers and workers with comparable expertise in private industry will remain a long-term phenomenon.[55] Newly developed incentive loan and tuition subsidy programs adopted in more than half of the states to attract college students to careers in mathematics and science teaching have thus far failed to pull in significant numbers of new recruits.[56] Despite the public attention devoted to the issue of shortages of qualified mathematics and science teachers, misassignment of untrained teachers to these courses has actually increased in the mid-1980s, partly as a consequence of upgraded academic graduation requirements in those subjects. Many other forces are at work in schools and in the society that may blunt or halt reform efforts. At this point, we do not know whether the 1980s will be recorded as an era of modest changes or a real historic turning point for American schools.

5 Forging a Partnership Between Education and Industry

Business has now been called on to help rescue the American schools. In the name of improving the caliber of the future work force, industry groups have become involved in a host of cooperative programs with the public schools. Such collaboration is seen as a partial aid for schools' fiscal, curricular, and management problems. But these programs have to contend with the major obstacles endemic to joint efforts between institutions, and few of them will survive to have a lasting impact on school practices. Morever, despite the efforts of a few business leaders and industry groups in support of increased public funding for public schools, the general indifference of the corporate sector to the fiscal needs of schools will retard efforts to bring about a renaissance in American education.

A number of cooperative school programs have been developed in the late 1970s and 1980s. These include "adopt-a-school" efforts, expanded career awareness programs, stepped up donations of equipment and other resources to schools, closer employer cooperation in schools' job training efforts for hard-to-employ youth, establishment of industry-education councils, and greater business involvement in the politics of educational reform. Governmental reorganization of job training programs and vocational education have also strengthened the hand of business in setting policy in these areas. Interest in computer training has provided a point of contact and collaboration between schools and high technology companies. The U.S. Department of Education, which has strongly promoted the partnership concept during the Reagan administration, estimated more than 30,000 school-business partnerships nationwide were operating by 1984, a notable upsurge in such joint efforts. (The term "partnership" will be used broadly in this discussion to include any program that involves business in the public schools, even though a truly reciprocal, equitable relationship between the two rarely exists.)

Business influence on public education and concrete ties between the two were significant in the late 1800s and first two decades of the twentieth century, when public education was in the process of becoming a mass institution. The development of vocational education, testing, ability grouping, and the adoption of "scientific management" in school organizations were all supported and actively encouraged by big manufacturing interests. A recent generation of revisionist educational historians and economists have described and analyzed the important role that big business played in shaping the ideology, curriculum, and organization of schooling through the formative years of the American elementary and high school.[1] From the 1930s through the 1950s, large manufacturing interests played a less direct role in influencing educational policies, content to leave school policies in the hands of school administrators and local school boards that had substantial small business representation on them.

But school-business ties and mutual interests eroded during the 1960s and 1970s. Educational institutions were given a host of new responsibilities as previously dispossessed groups demanded more equitable treatment. Desegregation, bilingual education, special needs education, and remedial programs for low income groups, among others, reflected the needs of newly organized citizen constituencies. Teachers formed active collective bargaining groups and students too became a vociferous interest group during this period. A wave of antibusiness and antiestablishment sentiment, fueled by American participation in the Vietnam War, the Watergate scandal, and protests against racism and sexism, swept college campuses and a significant segment of the adult population, leaving its mark as well on public school curricula and organization. Many teachers and pupils became hostile to business philosophy and conservative political outlook, and big business itself withdrew from some of the active ties it had once had with school programs.

Beginning in the late 1970s and continuing through the 1980s, however, business-education activity began to resume again. Some of the reasons for this renewal of interest in revitalizing such links have been outlined by Michael Timpane, president of Teachers College at Columbia University: employers' increased recognition that students need a broad-based academic curriculum in order to cope with changing work competencies; employers' growing concern over increasing deficits in basic skills among new employees, necessitating expensive corporate remedial programs; students' rising demands for career-related training; the long-standing and increasing rates of youth unemployment, highlighting the need for a better school-to-work transition; and an awareness among current school administrators that in an era of shrinking resources and citizen interest, business could provide political support for public education.[2] The general lessening of an antibusiness climate in American society beginning in the second half of the 1970s also contributed to the growing rapprochement between industry and the public schools.

It was in the early to mid-1980s, however, that highly publicized calls of public and private policymakers for closer collaboration between the two institutions occurred. President Ronald Reagan and his secretary of education, Terrel Bell, spoke strongly in behalf of corporate involvement in public schools as a way of compensating for diminished public financial support. Such efforts fit the conservative view that voluntary private contributions of various kinds are preferable to a large or expanding government role. Other groups stressed corporate support as well.[3] For example, the blue-ribbon Task Force on Education for Economic Growth, a group of governors, business leaders, and educators organized under the aegis of the Education Commission of the States and chaired by Governor James Hunt of North Carolina, recommended in 1983 a whole series of partnerships between business and schools. These included encouragement of firms to mobilize resources for public schools, to offer management expertise, and to articulate clearly what skills are essential for employees.[4] Many states have since set up special commissions, which included business representation, to study and make recommendations for statewide school reform.

Of all the kinds of school-business collaboration developed in the late 1970s and 1980s, two forms stand out as most significant. One is the increased role of business in government-mandated education and training policy councils. The other is the more visible participation of some companies and business associations in the politics of educational reform and funding at the state level. Several pieces of federal legislation passed in the 1980s have restructured policy groups so that business has a dominant voice. The Job Training Partnership Act (JTPA), which replaced CETA training programs nationwide in 1983, required that local Private Industry Councils (PICs), groups with a majority business representation but also including labor, education, and government representatives, determine the content and delivery of job training programs for low-income people. Some of the money goes to job training for teenagers and necessitates a cooperative relationship between the schools providing such training and the local PIC. The hope is that as PICs become well established in a community, they can become the vehicle for other kinds of collaborative relationships between companies and community institutions, including schools.

In 1984, Congress passed the Carl D. Perkins Vocational Education Act, which contained several provisions institutionalizing extensive business involvement in the governance of federally funded occupational education. This new Act was strongly supported by a coalition of business groups. The Math/Science bill passed in 1984 by the U.S. Congress also included provisions for school-business joint projects. And the National Science Foundation now stresses the importance of cooperative industry-education programs in the precollege science education projects it funds. The structured incorporation of employers in educational programs such as these guarantees a more permanent type of school-business partnership than has been common in the past. Institutionalized third-party organizations such as the PICs and state vocational education councils may ultimately provide the most fruitful, long-term association between education and industry.

Corporate influence in state efforts to upgrade public education is the other significant type of school-business collaboration to emerge in the 1980s. One of the most important examples of business involvement in schools came from the California Roundtable, the most influential business organization in that state. The Roundtable conducted a far-reaching assessment of the condition of California public education, publicized its findings and proposals for change, and worked actively in the political arena to bring about some of those changes. The recommendations called for the raising of educational standards, the upgrading of technical education, the reforming of school finance, and the strengthening of the teaching profession.[5] The Roundtable essentially concluded that business should become a political ally of public education as long as educators and public leaders were willing to pursue educational reforms. This kind of collaboration was probably the most meaningful of all school-industry ties being proposed at that time since it helped to lead to some concrete statewide reforms that were enacted into law in 1983. Prior to this time, business was not a significant actor in the formulation of state school policy. Moreover, the activities of the California Roundtable encouraged similar involvement by Roundtables and business groups in other states, such as Washington, Minnesota, and Massachusetts.

However, despite the increased number of successful collaborative ventures and the growing call for cooperation between public schools and companies, the two institutions still have very little to do with one another. Information gathered from personal interviews with corporate managers and educators on Route 128 and in Silicon Valley bear this out. Companies are much more likely to have ties with colleges and universities than they are with public schools. High technology firms have special characteristics that make their relations with schools particularly problematic. While a number of model programs have been created, especially in the area of computer education, numerous barriers exist between public schools and these companies. The presence of these obstacles inhibits the development of sustained and meaningful ties. Inflated expectations of some policymakers have failed to take account of the difficulties inherent in trying to build these relationships.

Barriers to Partnerships

COMPLAINTS ABOUT ONE ANOTHER

There has been long-standing mutual mistrust between educational institutions and American businesses. The high technology firms and schools in Silicon Valley and the Route 128 area in Massachusetts are no exception to this general phenomenon. Educators have long criticized industry for being too shortsighted and too self-interested when they become involved in educational matters. They often claim that companies are only concerned with immediate profits and not long-term social concerns. Teachers and administrators accuse industry of wanting to turn secondary schools into narrow technical trade institutes with little regard for the necessity of a broad educational experience. These attitudes, long held by educators, were expressed in interviews conducted as part of this study in California and Massachusetts with teachers and other school officials:

"I wouldn't count on industry for anything. They are just out to make money." A Silicon Valley vocational education administrator

"Gifts from companies come with strings attached. They would use the entree to a school as a secret recruiting mission. . . . You can't be involved in cooperative programs with people you don't trust." An associate superintendent of a Route 128 public school system

"The companies are shortsighted in the long run. Profit is what matters to them. They always use the scare tactic of threatening to move out of state. Business makes donations to use for public relations purposes or for a tax break or to tie us into their equipment. They always want some reciprocation from us if they give us a grant, like being forced to buy their equipment. Route 128 occupational education instructor

"Industry is self-interested and narrow and will milk a situation to their own short-term advantage. Our product is people and developing kids to their fullest potential. We don't want a special interest group coming in and having a different goal and measuring outcomes. Chair of a mathematics department of a Route 128 secondary school

"High tech companies are very self-serving. Their only interest with schools is to produce manpower. They want public money for training for their industry." Director of a Boston area adult public training program

"High tech companies are efficient at what they do but they don't understand the educational process. They would like us to run an employment and training system with no liberal arts. They will criticize us forever because we train the whole person. They see the person as the extension of the machine." A Massachusetts state vocational education administrator

Interviews with industry managers in this research, however, revealed that educators' views of them today are based on something of a misconception. Almost all of the high technology managers who were interviewed spoke of the necessity for a broadly based academic education at the secondary school level. Industry executives today have moved in the direction of supporting an academically oriented school curriculum for students in all ability groups and tracks, one that stresses communication skills, mathematics and science, and problem-solving abilities. They realize that job skills change so fast today that what is needed is a flexible worker who has the background to learn quickly. A 1983 conference of educators and industry executives sponsored by the Santa Clara County Industry-Education Council concluded that the lack of permanence in job requirements meant that public schools had to educate people who could be retrained with relative ease. Employers no longer expect schools to provide state-of-the-art technical skills since it would be impossible for schools to keep up with technological changes and since much of what students would learn would be obsolete by the time they were employed. Most employers are content to take on specific, narrow training skills themselves, and are content if secondary schools give students a rigorous educational background in either the standard academic or vocationally oriented program of study. A 1981 interview survey of twenty-eight Silicon Valley electronics company managers by Joseph Bellenger, conducted under the auspices of the Santa Clara County Office of Education, summarized industry's perspective this way:

Students don't know the basics. They can't fill out an application form properly, they can't spell, they are poor at mathematics, they aren't good at interviews, they don't know what it means to work. What we want is a well-rounded individual who knows the basics and appreciates the humanities; who has an understanding of self, others, and the business, cultural and political elements of our society, with emphasis on what is required to make the free enterprise system function and earn a profit.[6]

Thus, as job skills change more rapidly, employers emphasize the value of a broad education. In this respect, they are no different from other interest groups in the society.

Business managers have always complained that new workers are inadequately educated in school, but the intensity of their complaints (as well as those of college professors teaching freshmen) has risen in recent years. Their complaints focus not only on deficits in academic preparation but on the lack of the work ethic as well. The views expressed by the director of training at one of the largest high technology firms in Silicon Valley typify industry attitudes:

So many degrees today are worthless. We have to pick up some of the slack and make up for training deficiencies. We've watered down our internal training courses to push people through. The high school graduate functions at an eighth grade level. We want a person who really *is* what their degree says. If they have an Associate's degree, they should have competencies at that level. And younger employees lack commitment and responsibility. They should see a job through at least for a year.

A vice president for human relations at a Silicon Valley semiconductor firm put it this way:

There are problems with both attitude and cognitive skills of new employees. The work ethic of yesteryear is gone. Younger people are more concerned with quality of life issues. Prospective employees, even engineers, ask about recreation programs and sabbaticals before they're even hired.

He complained as well about students' lack of "economic education," long a frequent complaint among businessmen from all sectors:

Children and adults in our society don't understand the basic elements of our economy. Graduates of our local schools consider profit a dirty word. They don't understand profit. They have no idea of capital formation, return on investment, competition and the impact of legislative elements. . . . The excitement of a business career should be stressed in our schools as well as business' relation to the nation's growth.

Many executives cited the school experiences of their own children (most of whom are in the public schools contrary to what many people think), in discussing weaknesses in the educational system. Corporate criticism of the schools centered on beliefs that low academic standards and expectations produced mediocrity. Many managers felt that "excellence" had been sacrificed in the effort to reach children of all ability levels and backgrounds. The Silicon Valley businessmen were generally more critical of the schools there than were the Boston area executives. The latter tended to point out that suburban Boston schools were fairly good while the Boston public school system was exceptionally weak. Still, even there, the executives articulated critical views. As one Route 128 chief executive officer put it, "My children are in one of the best public school systems in this area, and if that's the best, it makes me wonder what the rest are like." Several bemoaned the fact that their children were uninterested in mathematics and science. Educators are sensitive and defensive about industry criticisms of the quality of instruction in their schools, a point that came out in a number of the interviews:

"Educators don't want to hear how wonderful the job opportunities are and how lousy the kids are. High tech firms always tell us how lousy our product is." Associate superintendent of a Route 128 public school system

"From industry's perspective, we only graduate functional illiterates. Their literature is offensive about our programs." Associate superintendent of a Route 128 public school system

"Industry thinks we are monumentally screwing up. They think they could do it better. They feel we are losers and we don't quite know what we are doing. Our image is low with them." Associate superintendent of a Silicon Valley public school system

DIFFERENCES IN VALUES AND AMBIANCE

There is a clash in values between the strike-it-rich mentality found in many high technology firms, particularly in Silicon Valley, and the more service-oriented human relations perspective characteristic of school personnel. While it is easy to exaggerate this difference, it nevertheless remains as an impediment to cooperation. The "nouveau riche cockiness" of some high technology entrepreneurs irritates public sector officials.[7] The competitive, frenetic, high-pressured business environment of many of these firms is a profoundly different work culture than that experienced by educational personnel, which makes it difficult for them to understand one another. Some of the company managers interviewed described the atmosphere on the job:

"The top corporate officials in this company are vociferous, vocal people. It is a hard pushing, hard driving company with a lot of high powered pushy people with big egos." Training director of a Silicon Valley electronics firm

"Is there a higher divorce rate in the Valley? I'm not sure but I've been on the verge of divorce ever since I came to work here. Silicon Valley is a real pressure cooker. If you lose, you lose your job. If you win, you make a lot of money." Manager at a Silicon Valley semiconductor firm

"Companies here are always looking for an immediate payoff and are unusually greedy and self-protective. High technology companies are much more paranoid about trade secrets and obstinate about school involvement." Manager of a Silicon Valley semiconductor firm who is active in school projects

It is difficult for schools and high technology companies to interact productively because the two institutions are at such different points in their history: schools are declining in enrollment, test scores, and financial support, while high technology firms are the current success story of the American economy. With one institution on the upswing and the other "in a dismantling mode," as one educator put it, it is hardly an atmosphere conducive to harmonious relations. True partnerships are difficult to establish when one of the partners is perceived as being more powerful and more successful than the other. A recent article in the bulletin of the Harvard Graduate School of Education contained a vignette that illustrates the high status accorded high technology companies and educators' own doubts about their self-worth. The author, educator Roland Barth, describes an incident at a party he attended:

The . . . party was attended by some of the most able school principals in the Boston area. Over in a corner of the room someone asked a stranger to the group—the husband of one of the principals—what he did for a living. He answered that he was a personnel recruiter for a high technology company in Boston. The conversation in the entire room abruptly ceased as everyone looked with renewed interest at this man. E. F. Hutton could have done no better.[8]

DIFFERENCES IN MANAGEMENT AND POLITICAL PHILOSOPHY

Electronics industry officials in both Silicon Valley and the Boston area believe that deficiencies in public schools are brought on more by poor management rather than by inadequate funding. They feel that more money will not necessarily lead to better education. "You can't throw money at education," argued one minicomputer company manager who spoke disparagingly about his brother's experience tie-dying T-shirts at a local suburban Boston high school. One administrator of collaborative industry-education projects in Silicon Valley claimed that local business people still did not fully appreciate the financial difficulties of the public schools, even in 1983 when the full impact of Proposition 13 had been felt in the California public schools and when the financial condition of the educational system had received a great deal of publicity: "As bad off as the schools are, I'm not sure the taxpayers and the business community understand the depth of the schools' fiscal problems. It is hard to get a corporate leader to understand the cataclysmic difficulties of the public schools."

Industry officials are not particularly sympathetic to the plight of school systems that have experienced budget cuts or to the condition of laid off schoolteachers since layoffs are such an accepted part of the American corporate environment. "Schools moan about budget cuts but we have to live with that all the time," explained the vice president for human resources of a major Route 128 electronics firm. A manager of a Massachusetts minicomputer firm put it this way: "Our corporate budgets get cut all the time. Teachers complain about cuts but these reductions are a fact and you can't cry over beer you spilled yesterday. You have to get the job done and quit crying about it." An educator in Silicon Valley who has worked with many firms in industry-school projects during the recession of the early 1980s described industry's attitude as one of "we're managing our problems, why can't the schools?"

Still, by 1983, Silicon Valley business executives had become more aware of the need for better funding of the public schools. But the general tenor of the business community there was that the schools were not just undersupported but nonproductive as well. As a result, business support has been contingent on school reforms of various kinds. Most of the corporate officials interviewed on the two coasts cited the need for merit pay for teachers, the abolition of tenure, and differential salary schedules in order to attract mathematics and science teachers (all of which teachers' unions oppose). They also want better school management, and the creation of scientifically and technically oriented magnet schools. Again and again they singled out the existence of teachers' unions as an obstacle to good school management—not surprising since the electronics industry itself is largely nonunionized. Industry's objections to educators' style and views were summarized by various observers in the two regions:

"The education establishment is perceived as intransigent in making changes. They are so defensive about change and unwilling to change on issues such as testing, tenure and teacher competence that their credibility suffers a lot." A Massachusetts state legislator active in state educational policies

"Educators need to learn how to listen. They see themselves as sophisticated people and have a certain arrogance. They are not good at listening and seeing another view. But now they are a little more humble because education is in such a desperate situation." A coordinator of industry-education projects in Silicon Valley

"There should be more public sector mentality fused into the private sector and more private sector mentality fused into the public sector. Schools need to make curricular changes such as eliminating outmoded vocational programs like cosmetology. School people don't realize what an important function work is in an individal's life—work is central for everybody." A Boston-area coordinator of a public adult electronics education program

The kinds of solutions to school problems recommended by industry, such as merit pay and a more rigorous evaluation of teacher performance, are not popular among most educators. By and large, teachers and administrators stress the need for more resources for schools while businessmen emphasize the importance of alterations in school organization and management. Thus, opposing views on school reform efforts constitute another barrier between companies and dominant segments within the educational community.

An underlying and fundamental difference in philosophy on the role of government exists between industry officials and educators. The executives interviewed in Silicon Valley and Route 128 believe in the conservative view that government, particularly at the federal level, should be limited in its scope and function. This "small government" philosophy of business causes them to oppose new government spending programs in domestic areas, including public education. An electronics executive in Silicon Valley, for example, explained that "companies are ambivalent about K-12 public school policy. Managers are entrepreneurial and Republican and believe that government should be small, decentralized and relatively non-bureaucratic." Another Valley executive, active in programs with local schools, pointed out that "companies are in a paradoxical position. We want to reduce the size of government so we can't at the same time support more money for education, even if that would help the company in the long run." A news article on a 1983 conference of semiconductor executives in San Diego reported that "at the mention of a larger government influencing the economy, the audience reacted with suspicious questions about the current high tax burden and government incompetence."[9] The executives applauded a speaker who "argued that government policy is safer if it

remains 'incoherent and chaotic.'"[10] It is not surprising, then, that a
1984 poll conducted by the *San Jose Mercury News* found that 77
percent of high technology executives were planning to vote for Ronald Reagan in his bid for reelection to the presidency.

Industry attitudes about government incompetence were
reflected in the statement of George S. Kariotis, founder and chairman of the board of a Route 128 high technology firm, as he left his
position as secretary of economic affairs for the state of Massachusetts in 1982:

There are a hell of a lot of very bright people working for government. We
thought we had them all in the private sector. I was shocked by how hard
people work. At the management level, people work harder than in the
private sector, and for lousy pay. I'm much more appreciative of the effort
that goes into this government scene. I respect it now.[11]

Educators, of course, espouse relatively liberal views on the role of
government. They believe that generous financing of schooling at all
levels of government is crucial to the adequate functioning of the
educational system, and think more highly of the competence of
public sector employees.

PROPOSITION 2½: THE TWO SECTORS
IN COMBAT

Relations between the high technology companies in the Route
128 area and the local public schools, never strong to begin with,
became more strained during the early 1980s. Public school administrators and teachers, particularly those who were politically
knowledgeable, were angry at the support that the Massachusetts
High Technology Council gave in 1980 to Proposition 2½, a property tax-cutting measure. Indeed, approximately half the companies
in the Council, the major trade association for the electronics industry in the state, donated a total of $229,000 to the Proposition
2½ campaign, 60 percent of the total amount spent on behalf of its
passage.[12] (By contrast, the electronics industry did not take sides
during the campaign for Proposition 13 in California in 1978.)
Council officials acknowledge that without their last minute fund-
raising, the tax-cutting initiative would not have passed.[13] Industry
executives believed that high taxes were a barrier to the recruitment
of desperately needed engineers to the Boston area and they sup-

ported Proposition 2½ mainly for this reason. They were not intent on reducing resources for education, believing instead that budget cuts would only eliminate waste and inefficiency at the local level.

Resentment among educators against the High Technology Council's support for Proposition 2½ was high, especially in those districts that had experienced significant cuts in their budgets. The resentment continued long after 1980 in part because the Council showed no sign of regret for their support of the measure, even when it became apparent that some school systems were hit very hard by the reduced financial support, and because the Council persisted in opposing some modifications of 2½ that would have eased its impact on local budgets. Comments from some of the educators interviewed illustrate the depth of their feeling on this issue:

"Around here high tech is a nasty word." A science department chair

"I have a terrible hatred for high tech." A mathematics department chair

"I won't sit in the same room with those people." An associate superintendent

"High tech people are not too smart on effects on human beings. They are not sensitive to the morale issues of teachers. They should teach for a week and *then* talk about morale." Superintendent of a regional-vocational technical high school

One industry leader acknowledged that the High Technology Council had a "black knight image" and had done little to ward off the backlash. Only two of the public school staff interviewed had supported Proposition 2½. Officials in the one town whose budget was unaffected by Proposition 2½ expressed little resentment against high technology companies, but they were the exception.

Over and over again, educators who were interviewed repeated that they could not understand why firms, which depended on an educated labor force, would support a measure the school people saw as detrimental to public education. "I can't understand their attitude," said the chair of a science department. "Businessmen *know* money is what makes things happen. You have to invest in a saleable product. I don't know what these people want." Many believed that the firms must not have realized what the ramifications of Proposition 2½ would be when they supported it. And the great majority claimed that the Massachusetts High Technology Council would come to regret its actions. Several school administrators argued that company officials did not understand the importance of teachers' morale in the educational process. "Morale is crucial when you can't turn your heat on until December 1st because your pay is so low," commented one associate superintendent. "Two and a half's symbolism is what is important," he added. "It tells teachers they're bad and they don't get their allowance."

The attitudes of the high technology officials interviewed were worlds apart from those of educators when it came to assessments of the impact of Proposition 2½ on the public schools. Three-fourths of the managers expressed the view that the budget cuts would fall only on peripheral programs ("mystery novel" courses, for example) and would not have an impact on such core academic courses as mathematics and science. Many said the cuts would strengthen the academic core because it would force systems to be "more focused" with priority given to scholarly subjects. Others said it would allow some communities to benefit from a "new spirit of volunteerism" or would eliminate "layers of unneeded management in schools" and other forms of waste. Most felt that if academic programs were cut, it would be the result of politically motivated mismanagement and not due to real fiscal shortfalls. A minority believed that the budget reductions would harm solid academic programs, a process that they claimed would ultimately cause Proposition 2½ to backfire on industry. Several bemoaned the losses of music and the arts, not only for their intrinsic value but also because they saw a correlation between music and interest in software engineering.

The financial crisis and curricular problems confronting the California public schools in the first half of the 1980s had the effect of raising business consciousness about the needs of schools, and, in a variety of ways, to be discussed later, industry statewide became a supporter of increased funding. Although Silicon Valley electronics executives were generally more critical of the schools than were their counterparts on Route 128, they were more willing to become political allies of schools at the state level by 1983. Of course, this alliance developed in part because absolute and relative funding for the schools had fallen so low in California during the 1970s and early 1980s. Because of this developing awareness and alliance, industry-school relations in the Valley improved somewhat during the 1980s.

ORGANIZATIONAL BARRIERS

Aside from the specific and transient issue of high technology industry's support for Proposition 2½ in Massachusetts, more fundamental organizational conditions of the companies make it difficult for them to interact productively with the public schools. Barriers between the two institutions are generally more numerous in new, high growth companies than in mature firms. The very dynamism, innovation, and rapid growth that characterize most high technology firms undermines attempts at extra-institutional ties. These companies, some of whom have growth rates exceeding 30 percent a year, operate in a competitive environment under the intense continuing pressure to get new products into the market. One company studied, for example, claimed it developed a new product every twelve working days. And these products often become obsolete in less than two years.

The firms need to put their cash back into research and development and need to focus their energy on the central task of the company. Outreach to schools in this context, except by a handful of large older companies such as General Electric and Raytheon, is of rather peripheral concern. Most high technology firms are small (three-fourths of them in Massachusetts have fewer than a hundred employees) and relatively new and have not yet developed institutionalized corporate giving programs. Comments from managers interviewed in Silicon Valley and Route 128 reveal the depth of this phenomenon:

"High tech companies move so fast that they get caught up in business decisions and overlook schools."

"Executives in Silicon Valley are so busy working 24 hours a day that they have no time to relate to schools."

"We are running so fast we haven't had time to reflect on education."

Rapid technological change, frequent alterations in company organization (mergers, spin-offs, etc.), the unpredictability of work force needs, and the ever-present vicissitudes of the market lead to short-term planning cycles. Markets for many high technology products are unstable and the products themselves are still undergoing frequent changes. (The microwave industry, whose major customer is the U.S. government, is one of the few mature and stable segments of high technology industry.) Intense competition among makers of personal computers, for example, is leading to the inevitable shakeout in the industry, a shakeout that occurred much sooner than many expected. Some analysts predict that only ten or twenty out of several hundred companies manufacturing these computers in the mid-1980s will survive. Some will be swallowed up by larger firms while others will cease to exist in any form. Others predict that among the fifty robotics firms now operating, only twenty-five will continue to exist in the long run. Industrial giants such as General Motors, IBM, and Westinghouse will take over or force out many of the smaller manufacturers of robots. Moreover, two-thirds of the world's robots are already made in Japan. Many computer software firms are also falling victim to the competitive volatile conditions characterizing that market.

Even large established computer companies face formidable competition from IBM, AT&T, and foreign business firms, notably the Japanese. Digital Equipment, for example, saw its stock tumble from 100 to 67 in just three days in the fall of 1983, reducing the company's market value by over $2 billion. The headlines on the business page of the *San Jose Mercury News* on one day in October 1983 give a sense for the volatility of computer markets: "Atari reports huge loss"; "New loans approved for Osborne" (the Osborne Computer Corporation filed for bankruptcy in September 1983); "Victor Technologies scurries for funding" (another later bankruptcy); and "IBM's earnings rose 25% in 3rd quarter."[14] Many corporate leaders failed to anticipate such trends as the popularity of microcomputers and the needs of nontechnical users of computers. The semiconductor industry, described as manic-depressive, has gone through severe slumps (in the mid-1970s and early and mid-1980s) and alternating periods of tremendous growth. Most of the semiconductor manufacturers have been acquired by larger firms, and much of their market has become dominated by the Japanese.

Other business analysts predict that "unfriendly takeovers" of some high technology companies by cash-rich firms will begin to occur, further unsettling the environment within which the companies operate. Many "friendly takeovers" have already occurred. In a two-week period in 1983 alone, several such takeovers happened on Route 128: Itek Corporation was acquired by Litton Industries; Data Terminal was purchased by National Semiconductor; and Instrumentation Laboratory, Inc., was taken over by Allied Corporation. A long news analysis of the problems of Data General, a Boston area minicomputer maker whose fortunes plummeted in the early 1980s and then revived, concluded by quoting a computer industry analyst: "There's one thing that abounds at Data General and that is uncertainty."[15] Uncertainty, in fact, characterizes a great many high technology companies and makes long-term planning difficult.

DIFFERING TIME PERSPECTIVES
AND PERSONNEL POLICIES

The present-oriented planning perspective characteristic of American business, particularly high technology firms, contrasts sharply with the stability of the schools that tend to operate on five-year planning cycles. Such different temporal perspectives inhibit the development of viable ties between high technology industry and education. Too, Silicon Valley companies have developed a distinctive reputation for an informal and quickened pace of work, a style characteristic of some of the Route 128 firms as well. It is no surprise that executives in these businesses become frustrated with what they perceive as the glacial pace of educational change. Business people and administrators who run industry-education collaboratives commented repeatedly on this issue:

"Our biggest frustration is with the school's slow pace of change. The culture of high technology companies is one of fast action. We do things quickly. Schools, by contrast, are a maze of bureaucracy."

"Everything is in a change mode. Schools can't respond quickly to changing needs. Schools couldn't respond even when things were *not* quickly changing."

"Schools have an inbred vision of their role in the community, and they may be divorced from community needs. There are turf problems and there is compartmentalization at all levels. People don't communicate with each other. Education has never been known as an innovative institution."

Educators, for their part, defend their more deliberate pace. "We will never really keep up with industry [in vocational education] but that's not our function," explained one Silicon Valley industrial arts educator. An administrator of Massachusetts vocational programs statewide put it this way: "What companies need today we may not see tomorrow. Companies can start up, close, merge at the stroke of a pen. But we serve the public and don't have that capability and don't want it." A number of educators criticized companies' manpower projections since the firms tend to operate with such short planning cycles: "Any projections beyond six months is crystal ball gazing," commented one educational administator.

The short-term time perspective of high technology firms helps explain why they assign low priority to relations with the public educational system. Industry executives may say education is important but are not willing to assign it any great significance in company efforts since they hire few students directly out of high school. When companies are interested in concrete educational projects, they look to higher education. "Businesses don't get excited over a high school kid; instead they will be interested in a community college student. Working with high school students is a long-range investment," explained a Silicon Valley vocational education administrator.

High turnover or personnel in dynamic high technology companies ("the silicon shuffle") also hinders the development of long-term collaborative efforts between the firms and the schools. Employee turnover in electronics firms nationwide in 1982 was estimated to be 23 percent with Silicon Valley and Route 128 firms reporting higher figures.[16] During periods of rapid growth, turnover rates are even greater. "It's hard to relate to a body rotating through a position," lamented one educational administrator. But "turnover will never change here; that's the way industry is," observed a manager of a major Silicon Valley firm. Some industry people interviewed felt that the "tremendous body snatching" that goes on between companies and the firms' competition for product markets interfered with the industry acting in a unified way on educational and other social policies. This was particularly true during the late 1970s when growth rates for many high technology firms were extraordinarily high. "It's hard for companies to cooperate," commented the personnel director of a Silicon Valley semiconductor manufacturer. "There are many company spin-offs and people hardly speak to one another." Only in the late 1970s did companies begin to work together on a variety of issues with the formation of the Santa Clara County Manufacturing Group and the Massachusetts High Technology Council. Unified industry positions are now more common but still difficult to devise and put into action.

PRACTICAL DIFFICULTIES IN
IMPLEMENTING COOPERATIVE PROGRAMS

Even when company executives want to help solve the problems of the public schools by contributing resources to them, they face the problem of choosing an effective intervention strategy. Unlike engineering schools where specific problems are obvious and where corporate efforts can make an immediate difference, public schools are large amorphous institutions that are relatively impervious to change from small-scale intervention programs. "The problems are so big, so bureaucratic and complicated that you don't know where to begin," explained one industry trade association official. A Silicon Valley business executive echoed similar sentiments: "It's hard for industry to decide what button to push in dealing with the problems of public schools because there are few focal points to attack."

Moreover, managers in companies who do take an active interest in educational programs with the public schools are usually only marginally influential in their own company. They are not in a position to commit company resources to promising programs and are more vulnerable than others to having their budgets cut. When recessions hit, projects with the public schools are especially likely to be reduced. Managers interviewed frequently made comments like "we're still in the talking stage," or "we're still figuring out our role in secondary education," and both industry and education people admitted that promising programs founder as a result of mutual lack of follow-through.

One reason why cooperative programs between the companies and the schools never get off the ground or fizzle once started is that both institutions currently lack the personnel to assign to such efforts. Even small industry-education efforts take a long time to get started. "It takes several meetings to get a $45 bus for a field trip," complained one Route 128 educator. The administrator of numerous industry-education projects in California lamented the fact that it "requires so much energy to have a small project. It's so damn slow it's pathetic." Recessions and market uncertainties in the private sector and budget reductions in the public sector have meant that companies and schools do not have people available who can devote the large blocks of time needed for successful collaborative programs. This has been an especially acute problem for educators in California who felt the impact of budget cuts sooner than their counterparts in Massachusetts:

"Time is at a premium for educators now because we are so understaffed. Cutbacks make it hard to coordinate our own shop. There is no time left over for coordination with industry. So we cut back on that." An administrator of secondary occupational programs in Silicon Valley

"School staff don't have much time to work on collaborative efforts with industry. We had 13–17 curriculum specialists in this office before—now only two are left." Associate superintendent of a large Silicon Valley public school system

"Some school systems here have reached a point in their decline where they are operating below a daily maintenance level and cannot do outreach to other institutions. They can only deal with emergencies on a day to day basis. An administrator of industry-education projects in Silicon Valley

Also, school officials are unsure whom to approach in a company ("it is a mysterious process") and claim that contacts are piecemeal and happenstance, often dependent on a personal connection. A number of school-company ties have developed, for example, because a parent in a school who is employed in a local high technology firm initiates the contact with the firm. Because mutual suspicion so frequently exists between the two sectors, it is often necessary for a third-party group to organize a collaborative project if it is to succeed. "There are troubles just with groups who *do* want to work together," lamented one Silicon Valley educator who has worked for cooperative efforts. "You need a broker or negotiator because there is so much suspicion." And even when there are third-party groups, school officials and teachers may back off from company initiatives because they feel firms are invading their turf. In a 1983 speech to a conference on school-business partnerships, Roderick MacDougall, chairman of the Bank of New England, pointed out that for years educators in Boston had resisted cooperative efforts with industry organized by the Trilateral Council for Quality Education:

For years it was a one-way partnership at many schools, with teachers and principals fighting the business involvement. Attempts to sit down with school officials and try and work out ways in which a business could be helpful to a school often resulted in demands simply for money and complete rejection of any input from the businesses as to how that money would be spent. In some cases headmasters wouldn't even come to the table to talk until they had been ordered by the Superintendent.[17]

Thus, resistance to collaboration can come from both the public and private sectors.

Variations in Approaches to Industry-Education Cooperation Within Each Sector

This study of the firms and schools in the Route 128 and Silicon Valley areas found that these numerous barriers caused ties between

high technology firms and the public schools to be generally fragmentary, short-lived and fragile. Joseph Bellenger's survey of
industry-education relations in Silicon Valley reached the same
conclusion. "While there are specific instances of excellent and profitable relationships between industry and education, many relationships are incomplete, somewhat loose and only marginally satisfying."[18] Of course, as Bellenger noted, there are significant pockets of
collaboration between some high technology companies and certain
school systems. A few educators interviewed had very positive feelings about working with electronics firms. For example, the occupational education director at a Route 128 high school, who devoted
substantial amounts of time and effort to developing ties with industry, was able to build such ties by being persistent in finding the
right person to contact in a company and by becoming involved in
any relevant industry-education collaborative he heard about. He
felt that if one does find the right company contacts, "many industry
people will give of their own time and good things happen."

In fact, school officials, even those most hostile to high technology companies, are usually willing to engage in cooperative
programs with them if the right opportunity presents itself. Their attitudes about working with the firms are often profoundly ambivalent. One Boston area school superintendent, for example,
spoke of the companies' support for Proposition 2½ with great
resentment and enumerated other criticisms of them as well, but
then argued that "we must work together," particularly in the areas
of curriculum development and job placements for students. The
chair of the science department of a Route 128 high school, referring
to the electronics companies, said in one breath, "I'm mad at them
but I would work with them if it would help the students." Similarly, the chair of the mathematics department in a neighboring
school system said, "I get very angry when I think about high tech's
support for 2½ but if someone had a realistic proposal, we could
work together." As in any large organization, significant variations
among personnel in school systems exist on this issue. A top school
administrator may resist any contacts with the firms but, at the same
time, the vocational educators in that same school system have had
long-standing programs with individual high technology firms. Significant variations exist among school systems as well, with some being much more aggressive about developing ties with business and
others having almost no contact.

Companies also display considerable variation in attitudes and behavior on school matters both within and among themselves. There is no uniform policy on school relations. Half of the companies in the Massachusetts High Technology Council, for example, refused to contribute funds in support of the Proposition 2½ campaign. Businesses in general have always differed in the degree to which they contribute to community insitutions such as arts organizations, charitable associations, and schools, and their motives for such support are diverse as well. A California administrator of statewide industry-education projects characterized firms' motivations this way:

Companies are like people. They have a personality of their own. They have their own style. Some are narrow [in their motivations for giving], some are avant-garde and broadminded. Some want economic education in schools; others take a broader view and want an articulate, intelligent community of citizens; some want to milk a specific labor supply; some companies just join for public relations and give a few bucks to schools to embellish its community citizenship role. The private sector is very diverse in its commitment and sophistication.

A manager of one Boston area high technology firm spoke of the way in which educators should approach a company if it wanted the firm's assistance:

Social programs are always irrelevant to the company. When the program becomes work or a bother or expensive, it will go away because it has no substantive meaning to the organization. When schools come for help, they should follow these rules: (1) don't ask for money; (2) don't cry; (3) don't talk about moral obligation or social needs; and (4) have a specific program with specific goals in mind that has something in it for everybody. WIFOM—"what's in it for me"—should be the guideline. There has to be something in it for everybody—students, business, teachers, the school system. Companies want a short-term, tangible, measurable response when they put resources into a program. If a school system says the company will gain in the long run by having a richer pool of work force talent out there, that is too soft for the company.

Some firms will be interested in supplying materials to schools that can only be described as company propaganda. A few companies will, however, make contributions of various kinds to schools because of a larger vision of what makes a good society.

Successful Industry-Education Partnerships

Despite all the obstacles discussed here, there are a small but growing number of partnerships between high technology companies and public schools scattered around Silicon Valley and the Route 128 area in Massachusetts. The most common and oldest form of collaboration exists between companies and vocational-technical programs (either within comprehensive high schools or within separate technical-vocational schools). Employer involvement on advisory councils for these educational programs is mandated by law and usually consists of meetings where employers give advice on the organization and curriculum of specific programs and courses. The overall improvement in the climate of industry-education relations generally by the mid-1980s seems to have strengthened employer involvement on vocational advisory committees. Vocational educators are more aggressively seeking out industry advice and many businesses are responding positively. Most vocational educators feel that business involvement on the vocational advisory councils is useful particularly when a new program is being developed.

A recent study by Patricia Flynn found that such councils at the vocational high schools, community colleges, and other occupational training institutions were very important to training programs in the Lowell, Massachusetts, area. The advice on labor market trends of these boards proved to be much more accurate than official occupational projections and thus enabled administrators and students to develop programs more in tune with local job openings. She also found that co-op work-experience programs were successful in giving students access to training on new equipment and providing full-time employment opportunities upon graduation.[19]

In some instances, the firms represented on employer advisory boards donate equipment to the schools. Some of it is useful although a great deal of it is obsolete. ("For one good item, you have to take ten bad ones" lamented one Silicon Valley vocational administrator.) A number of firms in both regions studied provided work-experience opportunities for vocational education students. And in Silicon Valley, a few companies allow their facilities to be used for training by secondary vocational programs. According to one administrator, the use of such facilities has been crucial to the existence of several occupational training programs (including electronics) since the Regional Occupational Programs could not have afforded the necessary equipment.

The provisions of the Vocational Education Act of 1984 mandate even greater involvement of companies in vocational education at both the state and local levels, so there should be a long-term strengthening of the historically close relationship between employers and vocational educators. These provisions include the creation of state councils of vocational education, a majority of whose members must come from business, which advises state officials on spending priorities and on labor market trends; the establishment of technical committees in each state composed of employers and labor leaders to come up with specific information on occupational competencies; the development of joint industry-education training programs in high technology fields; the requirement that vocational education and JTPA state councils have some overlapping members and coordinate their policies; and the stipulation that local PICs have to review occupational education plans that are supported by federal funds.

Plant tours and career awareness efforts are also examples of long-standing industry-education interaction. Indeed, most business efforts in nonvocational schools center on tours and speakers. Another area where some companies have been active in the last two decades are programs that are targeted for dropout-prone and hard-to-employ youth, usually in the inner city. The school-to-work transition has provided a focal point for a number of industry-education collaboratives across the country. Schools need help with these students, and business-oriented programs (which may include concrete work experiences) sometimes succeed in motivating students who have failed in academic settings. Companies feel they need to intervene early in the lives of potential employees if they are to develop essential work skills and attitudes.

The computer has recently become a new vehicle for cooperation between high technology companies and public schools. As one computer education coordinator in Silicon Valley put it, "The computer is hot and is a unit people can really see. Firms can target funds visibly for computer literacy programs." It is easier for a company to contribute to a highly focused and concrete project involving computers than it is to become involved in a more general program like "adopt-a-school" where firms provide a variety of services to an individual school or school system. A relatively small set of computers, perhaps out of reach of a school budget but within the contribution capacity of a computer company, can service an entire school. Because the computer is new to most schools' curricula, industry involvement in establishing computer education programs does not invade well-worn turf.

By contrast, companies have not, by and large, begun collaborative programs whereby mathematics and science teachers are hired in the summers by industry, nor do loaned company personnel teach mathematics and science in the schools on a part-time basis or help with curriculum development. There would doubtless be much greater teacher resistance if companies attempted to become more involved in the shaping of the content or pedagogy of mathematics and science, fields that are large and established in the educational community. Thus it is not surprising that despite the calls from a variety of local and national groups, firms have yet to become involved with mathematics and science education in any significant way. Of the sixteen mathematics and science departments studied in Route 128 high schools, only one had significant contacts with high technology companies. These ties included a donation of equipment, provision for a few unpaid internships for science students, and an attempt by one company to hire science teachers during a sabbatical year or summer.

Extensive involvement in school-business projects has been limited to a fairly small number of high technology companies in the two areas studied. Large firms in banking, insurance, and the utilities have been much more likely over the years to support cooperative programs with the public schools. Of the high technology companies that are involved with schools, Hewlett-Packard in Silicon Valley and Digital Equipment in the Boston area are the two undisputed leaders in fostering industry-education ties in their respective regions. William Hewlett and David Packard have had a long-standing personal interest in education, and their interest has left its mark on company policy. David Packard, for instance, spent ten years on the Palo Alto Board of Education. Both men have established foundations that frequently contribute to educational projects. Unlike most companies, Hewlett-Packard has a number of full-time employees who devote substantial portions of their time to contact with public schools. It has lobbied for increased expenditures of state funds for education and was the only high technology company in the Valley to donate money in opposition to Proposition 13.

Among other projects, Hewlett-Packard has loaned personnel and given equipment in support of developing curricula in drafting and in computer and electronics technology. Students from some local high schools use company labs at night for computer training classes. Intensive career awareness programs at several area high schools have been organized by the company, and it has fostered "adopt-a-school" relationships with a few area schools. Company officials were instrumental in building the local industry-education council as well. In 1983, Hewlett-Packard donated $23 million worth of equipment and cash to American universities and schools.

Digital Equipment, the second-largest computer manufacturer in the nation with headquarters in Maynard, Massachusetts, is a much newer company than Hewlett-Packard (founded in 1957), but has a history of active involvement with the educational institutions at all levels. The company, which did not contribute funds for the Proposition 2½ campaign, has long been involved in a range of programs: scholarships for students, donations of equipment to schools; help with curriculum development, particularly in the area of computer literacy; and loaned personnel. For example, Digital donated substantial amounts of computer equipment, software, and technical assistance to the Boston public schools in 1983 and began a $1.1-million program of contributions to forty-six high schools in Massachusetts and southern New Hampshire in 1984. Other firms— IBM in Silicon Valley and Raytheon and Polaroid in the Boston area—have also had long-standing relationships with the public schools in a variety of programs. But it has been difficult to expand industry involvement in community outreach of various kinds beyond the handful of companies that have been active in the past. Electronics firms are gradually developing sustained and organized corporate giving programs and ties with schools but the process is very slow. The ups and downs of companies' fortunes, even among large manufacturers, impede long-range planning in this area. By and large, the attitude in most firms is "let Digital do it."

The Industry-Education Council of Santa Clara County (IEC) has been somewhat successful in involving companies in school-business partnerships. The IEC is a chapter of the Industry-Education Council of California founded in 1973 by large employers, such as the Bank of America and Pacific Telephone. The Council is a voluntary association made up of representatives from business, schools (K-12 and community colleges), labor, and government, and is funded by dues paid by member organizations, corporate contributions, and government grants. Most observers feel that the specific programs of the group, many of which center on the needs of low-income youth, have been successful. They also point out that IEC meetings provide a neutral ground where educators and industry representatives can discuss mutual concerns. In the early years of the organization, there was an atmosphere of mutual suspicion among the various interest groups that sometimes erupted into angry confrontations at meetings, but that hostility has diminished over the years. Now communications are much freer and IEC representatives are more forthright than they used to be. Managers who attend meetings are now much more sympathetic to the plight of the public schools.

IEC officials are the first to admit the difficulty of overcoming institutional inertia and cross-agency suspicions when forming a consortium to work on even a small, focused project. Some of these projects include a systematic program of plant tours for educators, involvement in a high school program ("Learning to Earn") that will help assure entry-level employability of secondary graduates, and the creation of a computer van that provides introductory instruction in computers to a number of schools in Santa Clara County. More recently, Council members have expressed interest in placing priority on developing a political alliance that would be active in lobbying for greater resources for public schools at the state government level. There is no organization in the Boston area equivalent to the Industry-Education Council in Santa Clara County that brings together educators and industry people on neutral turf in a sustained and systematic way. However, in 1983, Senator Paul Tsongas announced a plan to develop a nonprofit corporation to act as a clearinghouse to match up schools and firms wishing to work together. And the Massachusetts State Department of Education in conjunction with business groups (the Business Roundtable and the Associated Industries of Massachusetts) has organized conferences and published materials to facilitate school-business pairings.

The Massachusetts High Technology Council (MHTC), an industry trade association with considerable political clout in the state, has also spearheaded some initiatives with public schools. High Technology Forums, which brought together educators and industry people to work on career awareness and some curriculum help, were organized in several regions of the state in the early 1980s. The Council has also helped write the curriculum and served in an advisory capacity for state-funded computer programming courses for laid off schoolteachers. A Committee on Computer Literacy has been organized to coordinate companies' efforts to assist teachers and students in computer education. The first step of the effort began in 1983 when Cullinet Software (whose chief executive officer, John J. Cullinane, is chair of the MHTC Committee on Computer Literacy) began training groups of teachers at the company's expense in week-long courses at its own industrial education facility in Framingham, Massachusetts. The Committee hopes that other high technology firms will follow Cullinet's example. The president of the MHTC was on a statewide commission that advised the governor and legislature on a package of school reform measures in 1983. The main thrust of MHTC's educational efforts is, however, aimed at the college and university level.

Some of the more innovative and meaningful partnerships between the companies and the schools center on employment and training of low-income youth. Two projects, the Peninsula Academies in two high schools in San Mateo County (adjacent to Santa Clara County) and Project COFFEE in twelve public school systems to the west of Boston, are aimed at providing an alternative occupational education program in computers and electronics for "alienated disaffected secondary school age youth." The Peninsula Academies, modeled after similar programs in Philadelphia public schools, include a Computer Academy at Menlo-Atherton High School and an Electronics Academy at Sequoia High School in Redwood City. Only tenth-grade potential high school dropouts are admitted to these "schools-within-a-school" where for three years students get intensive training in either computers or electronics from loaned industry personnel. Summer jobs, individual mentors in industry, and guaranteed entry-level high technology jobs are available to students in the program.

Thus far, the 137 tenth graders (most of whom are minorities) who began the Peninsula Academies program in 1981 have demonstrated improved attendance and achievement and have raised their occupational and educational aspirations. The Academies were started by the Stanford Mid-Peninsula Urban Coalition of Palo Alto, a community service organization that is dominated by business representatives, and have been sustained by business contributions of various kinds. Firms such as Hewlett-Packard, Lockheed, Varian, and Ampex have donated equipment and loaned employees, and ten high technology companies guaranteed summer jobs to the first cohort of Academy students who finished the eleventh grade. About sixty Silicon Valley executives agreed to be "mentors" for this group as well.

Project COFFEE is similar in that it consists of technical training in computers and electronics for dropout-prone students. Founded by the Oxford public schools and the nearby French River Teachers Center, the project has substantial support from Digital Equipment and the U.S. Department of Education. Digital has provided a whole array of services and equipment for this highly successful program. The Oxford schools, Digital, and the French River Teachers Center also teamed up to develop a series of two-week summer computer training camps for 750 disadvantaged youth from several Massachusetts cities. The project received matching state funds. The Oxford-Digital connection has spun off a variety of partnership activities that have pulled in other educational institutions and government agencies as well. Another program that has received significant industry support is the Massachusetts Pre-Engineering Program for Minority Students (MASSPEP) whose goal is to increase black, Hispanic, and native American representation among college and university students in the fields of engineering and science. Begun in 1979 by the Urban League of Eastern Massachusetts, the project has provided special educational programs both inside and outside of school for several hundred junior high and high school students in Boston and Cambridge. Raytheon, Honeywell, and Polaroid have been among the main financial contributors to the effort.

Other collaborative programs are aimed at lower income and hard-to-employ youth. The Private Industry Council in Boston has been instrumental in developing the Boston Compact, a unique partnership between the public schools and the city's business community. The Compact is an agreement between the public school system and more than 300 business firms. The agreement specifies that the companies will offer summer and entry-level permanent jobs to Boston public high school graduates in return for verifiable improvement in the school system. Beginning in 1982, the school district promised to increase the number of graduates from Boston high schools by 5 percent a year (currently only about half of the ninth graders ultimately graduate), ensure that by 1986 all Boston seniors achieve minimum competency in reading and mathematics needed for employment, evaluate high school headmasters on the basis of their achievement of prescribed goals, and supply annual reports describing student attendance, achievement, and placement. For their part, companies who sign the Compact agreed to a 5-percent increase each year in the number of graduates who are placed in jobs.

By the fall of 1983, the Boston business community had more than met its side of the bargain. It had signed up 273 firms as parties to the Compact, more than the 200 it had set as a goal. Four hundred and fifteen graduates were hired in full-time entry-level jobs (40 percent of the approximately 1,000 graduates annually who are looking for employment), exceeding the goal of 400 jobs. Another 600 graduating seniors obtained full-time positions in 1984. Almost 1,200 students were hired in summer jobs in 1983, more than 1,500 were placed in 1984, and 2,000 were employed in 1985. Preliminary evaluations showed the great majority of students hired in summer and full-time jobs performed well in their new positions. Moreover, half of the 225 companies participating in the 1983 summer jobs program had never been involved in such a job placement effort before. Compact staff work closely with the school system in developing programs to improve school performance. An additional agreement has been signed by twenty-five area colleges and universities who promise to increase the number of Boston public school graduates admitted to their institutions (along with financial and academic support) in exchange for measurable improvements in the high school retention rate and academic achievement.

In effect, the Compact means that the Boston business community has become a consciously active constituent group pushing the school system to improve levels of performance and accountability. Business leaders reasoned that since so few adults had children in the public schools in the city and since concern for the public schools had waned among other interest groups, it was essential that they step in and work for school change. This involvement by industry is motivated by two principal factors. First is the "business climate" of the city: according to Jim Darr, executive director of the Boston PIC, "It's not a good business climate to have your sparkling tower downtown when you've got public squalor a mile away."[20] Second, employers are worried about the long-term labor supply since a shortage of entry-level young workers is forecast for the late 1980s. Declines in the birthrate mean that business people will be forced to be less selective in hiring so it is in their self-interest to upgrade the quality of education of all students who will form the future labor pool in the city.

It is too early to assess the long-term effectiveness of this effort, but in the short run it is clearly a success. The dropout rate for juniors and seniors in the Boston high schools has gone down, influenced in part by the prospects for Compact-provided employment after graduation. It is unusual to have business involved in such a comprehensive school improvement program, particularly one that promises jobs as a reward for school success. Many educators and community leaders who were initially cynical about the program have developed a more favorable view of it. But high technology firms, with the exception of Digital, have not been in the forefront of the Compact, in part because so few of them are located in the city of Boston. Fewer than ten high technology companies have signed the Compact.

There are other important examples of school-business ties in Boston. The Trilateral Council for Quality Education has coordinated industry-education projects in the city since 1974, including the school-business pairings that developed as a result of the recommendations of the 1974 federal court order desegregating the Boston public schools. Some of the more successful efforts include partnerships between Gillette and South Boston High School, John Hancock Mutual Life Insurance Company and Boston English High School, and the Bank of Boston and Hyde Park High School. Programs of this kind generally focus on improving students' job-related skills and attitudes.

Several private sector institutions have made precedent-setting financial gifts to the Boston public schools in 1984 and 1985. The Bank of Boston announced a gift of $1.5 million for an endowment fund, the largest contribution ever given by a firm to a public school system in the United States. The earnings from this endowment are being used to finance innovative projects proposed by local school personnel, parents, and students. An endowment of $1 million to support intramural athletics and reading programs in Boston's middle schools was given by John Hancock. The Bank of New England began a five-year $300,000 program in 1984 focused on the continuing education of Boston public-school teachers, and New England

Mutual Life Insurance Company donated a $1 million endowment whose income will pay for college scholarships for graduates of Boston's public high schools. The size of these gifts and the fact that three of them take the form of endowments attest to the growing strength of school-business ties in Boston.

Electronics and computer companies are more active in computer education programs than they are in any other collaborative endeavor with public schools. Innovative partnerships of various kinds have sprung up in both Silicon Valley and the Route 128 area. Until 1984, when it was shut down, Raytheon Data Systems in Norwood, Massachusetts, ran the highly successful Project Access, which involved teaching programming to several hundred students in four high school districts. The company helped train students in routine coding tasks of software programming through school-based courses and then hired some of them as paid interns and part-time workers. A few graduates were hired full time by the company. Teachers from twelve local school systems were hired for several summers as part of the training program. The aim of the program was twofold: the company hoped to increase the long-run supply of programmers by encouraging high school students to major in engineering and computer science in college; and it wanted to reduce turnover of engineers by reassigning dull routine tasks to high school interns. The firm donated substantial amounts of equipment and time to the project, putting special emphasis on working with a cadre of teachers over a long period of time. The abrupt closing of this division of Raytheon, however, shows how vulnerable partnership efforts can be when only one firm is involved in the program.

Another program linking schools and high technology firms through the computer was begun in 1982 when the Mountain View–Los Altos High School District in Silicon Valley solicited support from Atari and Picodyne Corporation to provide a system that would link students' home computers with the school system's computers. Students can receive guidance counseling and school news over this hookup now and eventually will be able to get homework assignments, grade reports, and access to the library's card catalog. About 15 percent of the district's high school students now have a home computer but within a few years 90 percent are expected to own one. The total cost of $90,000, borne mostly by Atari, did not require any school district funds. This project, conceived by school superintendent Paul Sakamoto, has attracted national attention as a particularly original model of the practical uses of school computers. In another program, several firms in the Route 128 area, principally Raytheon, chipped in during 1983 to help establish a computer center for training and software information for a consortium of eighteen public school systems, the Education Collaborative for Greater Boston (EdCo).

The school-business project in Silicon Valley that has received the most publicity is the Institute of Computer Technology (ICT), which was established in 1982 in Sunnyvale and Saratoga. Initiated by the superintendents of the Los Gatos Joint Union High School District (which also includes Saratoga) and Fremont Union High School District (which includes Sunnyvale and Cupertino) and joined by the Sunnyvale elementary district, the Institute has been set up to offer courses in various computer-related subjects and electronics as well. High technology companies supported the effort because they believed such a school, in the long run, would help relieve the shortage of technical personnel available as potential employees. High housing costs in the area have convinced many firms that the Valley must "grow its own" labor supply since outsiders are reluctant to move there. Courses are offered at no cost in two sites after school and in the evening for public and private school students in the three school districts and adults in those communities as well. Operational costs are largely borne by the state of California but twenty-six local high technology companies have made substantial gifts of equipment and loaned personnel. What makes the project unique is that it is an independent legal entity, chartered by the state, that is jointly run by industry and the three public school districts. A board of seven members, five employed by high technology companies, governs the Institute. Industry has been remarkably cooperative and there is little friction between company representatives on the board and educators. Local high technology firms made contributions valued at over $500,000 within the first few months of the school's operation. Two executives, loaned from Hewlett-Packard and IBM, were full-time administrators of the Institute along with a school system official. By 1985, corporate contributions of various kinds totaled close to a million dollars.

Ironically, just as the Institute was undergoing a major expansion of courses, state funding for the project was cut off when Governor George Deukmejian vetoed portions of the public school budget in July 1983. The school was forced to cancel its summer courses, which had enrolled 400 students and on which $50,000 had already been spent. Local business people and educators in the Valley were astonished that an obviously successful model of industry-education collaboration could be closed down by the state, and they initially assumed that the governor must have inadvertently slashed the funds. However, the governor insisted that the cut was intentional and asked the participating school districts to pick up the financial burden. Since the latter lacked funds for the project, a campaign was mounted by those involved in the school to restore the money. As a result of pressure from the business and education community, as well as state legislative support, the governor restored $250,000 of the $1 million the school had originally expected. A similar financial crisis occurred again in 1985 when the Institute survived only after an emergency funding bill was passed and signed by the governor (who had vetoed a three-year financial package in 1984). Adequate funding for the next several years remains uncertain. In this case, it is the public sector that has been the weak and uncertain partner in the enterprise, not the business community as is often the case, demonstrating that the vicissitudes of government funding can destroy partnerships as well.

Another model of high technology firms' collaboration with schools in Silicon Valley is the computer van that tours schools under the auspices of the local Industry Education Council. With primary support from Atari, the computer van with its fifteen learning stations serviced 12,000 to 15,000 students during its first two years of operation and has offered a number of in-service training courses for teachers. The van's major mission is not to provide comprehensive computer literacy training but seeks instead to stimulate educators to plan for computer education programs of various kinds. Throughout California, high technology industry people serve on the advisory boards of the Teacher Education Centers set up in fifteen regions in 1982 to train teachers in computer uses.

California also adopted an "Apple bill" in 1982, which provided a generous tax credit for companies who donated computers to schools. The state law enabled Apple Computer and other companies to write off about 90 percent of the cost of manufacturing the machines. Apple donated one of its microcomputers to every accredited school in the state of California. IBM responded with a national giveaway plan of its own whereby it donated 1,500 personal computers to secondary schools in three states, including California.

The IBM effort is the most coordinated of the corporate donation schemes because it is tied in to teacher training centers. Fourteen high schools in Santa Clara County are receiving "class sets" of fifteen computers as a part of this program. IBM expanded this program in 1984 to twenty-six large urban schools systems across the United States, including Boston's public and parochial systems. Hewlett-Packard and other computer companies have also set up donation efforts. About 16 percent of the public school districts adopting computers in Massachusetts have received corporate support for such purchases.

While computer-related education is an obvious vehicle for collaboration between high technology companies and schools, educators still approach certain kinds of cooperation with caution. Many expressed the suspicion that companies involved in donation programs are not doing so with students' best interests in mind; instead, they point out, companies are using the programs as marketing devices with particular interest in reaching the home market. Schools that adopt computers from just one company may be left without repair services, new software, disk drives, and peripherals if a firm drops out of the computer market. Many schools, for example, bought Texas Instruments' microcomputers because of its student-oriented software capabilities (the first to run LOGO, a popular program in schools), but since the company announced its withdrawal of the product from the market in October 1983 these schools may face future difficulties with the machines on its hands and will not be able to purchase new ones as their needs expand. Further, most of the Texas Instruments' programs cannot be used on other machines.

Some educators interviewed expressed the view that, for the most part, there is little real overlap in the manner in which corporations and schools use computers. "The companies are highly sophisticated but is it the kind of skill that will fit in the schools?" asked one Silicon Valley educator. Many teachers and school administrators have become quite sophisticated in their knowledge of computers in education and do not feel they have much to gain from company expertise. The computer companies, for their part, did not aggressively pursue school markets initially, preferring instead to sell to corporate buyers. According to the regional marketing director for education of one computer company, "We are *responding* to the education market; we didn't create it. Schools are not really ready for technology but we can't just wait around. Are schools really ready for *anything?*" This manager expressed frustration with all the difficulties of selling to schools compared with the relative ease of selling to the corporate training market. Thus, some of the cynicism and suspicions that have generally characterized school-business relationships over the years sometimes surface in the computer education area as well.

Business Support for School Reform

One of the most important of the new initiatives between companies and schools is not in the area of specific industry-education projects but is instead the gradual emergence in the 1980s of business support for adequate funding of the schools and for school reforms of various kinds. The growth of this kind of alliance has become apparent in California and, to a lesser degree, in Massachusetts. This form of cooperation probably has more potential significance for the future of public education than any other kind of partnership. Even Albert Shanker, the president of the American Federation of Teachers, has spoken out strongly in favor of seeking out business leaders as allies in pushing for adequate education funding.

More companies, including some high technology firms, are actively lobbying state, local, and federal levels of government for increased funding for the public schools. Simultaneously, the business sector is demanding that increased funding be coupled with upgraded standards of student and teacher performance. Of course, the business community is not alone in its critique of American public education. Its view of what is wrong with the schools and what should be done to correct its ills is widely shared by the general public, a significant number of public officials, and many educators themselves. As Michael Timpane put it, after surveying business opinion of the schools, "The corporate interest in educational quality is, in sum, not distinctively 'corporate,' but simply part of a broad convergence of popular educational opinion."[21] The activities of the Boston Compact and the Industry-Education Council of Santa Clara County, and the interest of the Massachusetts Business Roundtable, the Massachusetts High Technology Council, and the New England Council (another business group) in public education are all signs of renewed business interest in educational change.

In California, two events occurred in 1982 that illustrate the kinds of political alliances being formed between industry and education. The first event was that business executives, including those in Silicon Valley high technology firms, strongly supported the succesful candidacy of William Honig when he ran for superintendent of public instruction against incumbent Wilson Riles. Honig won the support of the electronics community because of his emphasis on upgrading school standards, including a pledge to improve instruction in mathematics, science, and computers. David Packard, for example, threw his support and money to Honig after supporting Riles for many years. Another executive, David Bossen, the chief executive officer of Measurex Corporation, explained his support in this fashion:

Our raw material in the Valley is brain power. The declining test scores make us concerned that we are running out of our precious resource. We're not graduating enough engineers and scientists to support our continued growth. . . . I have made a campaign contribution I would not have made in the past. I have come to recognize the problem. It is not a one-year phenomenon.[22]

Similar views were echoed by Dr. Gordon Moore, the founder and chairman of Intel Corporation:

I'm not much of a political activist, but personally and politically, I have addressed my interest to the state superintendent's race. A lot can be done by setting high standards.[23]

The issues raised in this race increased the awareness of high technology executives of the crisis in curriculum and funding that was besetting California public education. At the same time, the California Roundtable conducted an extensive study of the condition of the state's public educational system and issued a comprehensive set of recommendations.[24] The report of this highly influential business group called for a longer school day and school year, increased academic course requirements for graduation, improved technical education, strengthened attendance and disciplinary measures, textbook upgrading, and mechanisms to strengthen the teaching profession. The group worked for passage of the Hughes-Hart Educational Reform and Finance Act (SB813), which contained these proposals and which passed in 1983. The executives followed through on their pledge that they would support increased monies to improve the schools' financial base if, at the same time, reforms of various kinds were passed as well. Prior to this period, business did not have a hand in the politics of educational change. Without this support from the business community, Governor George Deukmejian, a conservative Republican, would not have agreed to the increased school budget allocation that was passed.

Business involvement in the Massachusetts school reform movement has been more muted and more complex. The Massachusetts Business Roundtable formed an Education Task force that studied and began speaking out on policy proposals for public education in 1983. Its membership represents mostly older manufacturing interests as well as banking and insurance institutions, and, as in California, the organization is very influential. The Roundtable provided computer hardware and software to aid the state legislature's Joint Committee on Education in data collection on the condition of education in the state. Along with other business and education groups, it supported a series of regional conferences in the state in 1984 that provided school officials and business executives with information on how to establish a successful school-business partnership. The Roundtable also enunciated its priorities for school change to the Committee in 1984 when that body was considering a package of reforms. These priorities included support for academically tough statewide high school graduation requirements, more computer literacy and economic education, student competency testing, a longer school day and year, more homework, strengthened attendance and discipline laws, and competency testing and merit pay for teachers. According to banker Roderick MacDougall, head of the Roundtable's Task Force, when businessmen in the state were asked if they would support higher business taxes to improve public education, they answered "no" or "maybe":

The "definitely no's" explain their position by citing current per pupil costs vs. other states. . . . They feel strongly that current total resources being allocated to public education from local, state and federal sources are adequate to upgrade the quality through better management. . . . The "maybe's" are saying show me the legislative bills and price tag *together*. If the new programs eliminate excesses, bring in new efficiencies, and gurantee improved curriculum, an increased pace of academic learning and enforces standards among students and teachers, then some new tax dollars may be possible. If the new dollars will be simply to respond to teacher demands for across the board increases regardless of performance, then forget the business support. The "maybe's" then all become "no's."[25]

When an educational reform package reached a critical stage in the state legislature in late 1984, business groups, including the Roundtable, opposed the more expensive version of the proposals. Even though the bill included many provisions the Roundtable had previously supported, the organization balked when it became clear that a significant hike in teachers' salaries could lead to an increase in taxes. The opposition of this and other groups (including key state senate leaders and the governor) doomed the bill in that legislative session. Thus, an alliance between industry and teachers' groups, which had gradually evolved over a two-year period, unraveled when monetary concerns became paramount. This is not surprising since business groups, who are once again visible players in the politics of public education, have a deep-seated desire to keep public expenditure levels low. They may support reform measures and perhaps even a small tax increase to finance the implementation of the changes, but they will rarely fight for greatly increased budgets for domestic programs. Chronic underfunding of domestic public-sector budgets has become standard operating procedure in the United States.

Partnerships: A Summary

Important collaborative efforts between high technology firms and the public schools exist and are growing if Silicon Valley and the Route 128 area are any example. But it is misleading to expect widespread and sustained cooperative programs. While some firms have been able to make a long-term investment in the schools, most are so caught up in immediate corporate problems that public school issues do not command their attention. Both schools and electronics companies are concerned at this point with their own survival.[26] The public schools are struggling with shrinking fiscal resources and student enrollment, absorbed with school closings and possible bankruptcy. The companies, including some of the large ones, exist in a fiercely competitive and unpredictable economic environment that can bring about the sudden demise of even high flying firms. It is difficult to imagine much more than a "struggling relationship" given the context within which the two sectors operate.

Further, most high technology firms are located in suburban areas away from central cities, which means that they are relatively uninvolved in the current wave of corporate efforts to improve the employability of inner-city youth. The involvement of companies like Control Data, Honeywell, and Digital in such programs form the exceptions rather than the rule. As William C. Norris, founder of Control Data, has pointed out, the most important social contribution a company can make is to locate a plant in a central city.[27]

There is an underlying cynicism and despair in the commments of many who have long been active in efforts to bring companies and schools closer together. One administrator of such projects spoke of schools and businesses as "mismatched arenas" and described his efforts at collaboration over the years as "inconsequential." Another talked encouragingly about the progress that has been made in Silicon Valley in the last few years in industry-education efforts, but then ended by saying that he didn't know "how real the progress is." A Massachusetts administrator of a successful cooperative program discussed the project's success in garnering corporate funds, but later admitted that "it's a constant hustle to get money from industry."

Another problem common to most attempts at collaboration between schools and businesses is that successful partnership programs are difficult to replicate. Perhaps this is because the personality and leadership factors are so important in these efforts. It isn't the ideas that are so novel, it is the blend of commitment and ability of the leaders involved that is crucial to success. Since most cooperative programs are initiated by school officials, the leadership capacities of educational administrators are especially important. The economic and political context within which these partnerships operate varies considerably from place to place. Some regions, like Boston, have relatively bitter corporate–public school histories while others, in Texas for example, appear to have little political tension surrounding their relationships. The Southeast and Southwest have low percentages of workers in unions, including teachers, a factor that sometimes gives administrators a freer hand in arranging joint programs with industry. An area that has had a history of companies leaving in search of cheaper labor will have school personnel who are much more cynical about joint efforts with industry than regions where manufacturing is a relatively new sector of the economy.

Given all this, expectations for school-business partnerships should be modest, particularly partnerships with high technology firms.[28] But scattered examples of fruitful interaction will continue to occur and will probably increase somewhat in the near future.

There are some common elements generally found in successful partnerships.[29] Programs must be initiated by enthusiastic and capable individuals who will then lead the projects through a sequence of stages. The involvement of highly placed officials in both the firms and schools is also an essential ingredient of a successful partnership, and the program should also involve close relationships among many people in the participating company and school as well. This web of connections prevents a whole project from unraveling when an important figure in the program changes jobs or when company fortunes temporarily sag. It is also essential that adequate numbers of personnel be available to work on a cooperative program. Many have foundered because no one has been able to make it a priority and assign staff to work on it full time.

Public funds, beyond those available in a school's regular budget, are also needed in most industry-education projects. Few corporations can sustain long-term operating costs of a joint venture, and government financing is crucial for the maintenance of a smooth onging program that cushions the budgetary vicissitudes of corporations and schools. Officials working in cooperative projects need patience as it often takes up to two years to plan a program and another two or three years to see concrete results. It is important that participants in collaborative efforts have a certain amount of humility: information, suggestions, and support need to alternate back and forth between industry and education if a program is to be a truly cooperative one. Firms and schools appear to work best together when projects are concrete and focused with specific goals spelled out. The spearheading of a cooperative program by a third-party neutral group or a consortium of schools or companies also enhances the likelihood of success.

A consensus among scattered industry groups and business leaders and some educators is crystallizing on the solutions to the problems of public schools. Where the problems are dramatic, as in California, the city of Boston, and southern states, companies are becoming more involved in the politics of education. Their support for schools includes not only corporate willingness to advocate increased public funds when needed, but also a commitment to criticize and evaluate school performance. In some states, business support for school funding and reform will be crucial as it has been in California. In others, for example Massachusetts, industry involvement in educational change at the public school level will be important but probably not crucial. Indeed, business participation in the political process may result in a reduction in school resources, as was the case with high technology industry's support for Proposition 2½. In other states, companies are simply uninvolved in the politics of educational reform. In the long run, the most profitable partnership public education could develop with business is a political alliance based on an understanding of the essential needs of schools. After all, observed the chief executive officer of a Route 128 electronics firm, "marvelous things happen when people talk to each other."

6 How Higher Education Handles High Technology

Institutions of higher education have altered their programs to meet the changing realities of the microelectronics era considerably more than the public schools have been able to do. The goals, financing, and structure of post-secondary institutions of education differ markedly from those of the public schools, enabling them to be more responsive to shifts in their economic environments. Although industry executives criticize colleges and universities for moving slowly in changing their curricula, these institutions are much more fluid and flexible than elementary and secondary schools. Given the constraints on resources, higher education has moved remarkably fast in the 1980s to reorient its offerings in a more technical direction. Many states have reevaluated the relationship of their public higher education systems to economic development proposals and, as a result, have strengthened engineering, computer science, and other technical programs of study. Nevertheless, despite these efforts, colleges and universities have not been able to staff and equip new or expanded programs of study in an adequate manner. The long-term federal support that is crucial has not been forthcoming.

Higher education in the United States is a highly complex, differentiated system, and its response to changing economic climates is varied and uneven. Yet there is a surprising uniformity across the country with regard to trends in technical education. New schools of engineering and computer science are being established and old ones expanded; extension programs aimed at adult workers who wish to be retrained or upgraded have been developed; advanced research centers in electronics and computers are being created in all regions of the country; computer use by students has accelerated dramatically; and company-specific training for employees is increasingly being done by higher education institutions. Student enrollments in engineering and computer science as well as in two-year technical programs have jumped substantially in the 1980s. Governors view their public higher education systems as tools of economic development, and students see a technical degree as a ticket to secure and well-paid employment.

Colleges and universities can change more quickly and easily than public schools for a variety of reasons. To begin with, higher education has always had something of a vocational emphasis, training teachers, ministers, farmers, lawyers, and the like. When post-secondary education became available to large numbers of Americans following World War II, high schools became somewhat more removed from the immediate demands of the job market. Gradually, community colleges and four-year colleges have come to be seen as the crucial training grounds for the nation's future employees. Business leaders are now stressing the importance of a broad, academic secondary education and have shifted their concerns about vocational and career training up to higher levels of the educational system. Thus when economic forces change, it is colleges, universities, and other segments of the post-secondary training system (e.g., proprietary schools) that feel pressure to redirect their instructional programs to be congruent with these market trends.

Moreover, students have a great deal of latitude in their courses of study, unlike those in secondary schools who follow a more prescribed curriculum. Students are relatively aware of labor market trends and tend to select courses of study they think will lead to solid employment opportunities. Faculty feel that, if anything, students overreact to current employment trends in their haste to insure a marketable degree. College administrators may question students' long-run judgment in their curricular choices but feel nonetheless that they must respond to students' demands for access to certain programs. Unlike public school administrators, higher education officials must worry about attracting students to their institutions, which means they must expand or develop majors that are in great demand. This is especially true in an era of significant declines in the traditional college-age cohort. As one Boston area university administrator put it, "The University is a business. Students are banging on our door for engineering and computer science. There is such a huge demand, we *have* to respond to stay in business." Thus, competition among colleges and universities for students forces some schools to alter curricula. In some cases, institutional survival is at stake. Generally speaking, academic administrators appear to be eager to provide the kinds of programs students want while liberal arts faculty are decidedly less enthusiastic.

Other factors make higher education more responsive to work force trends than public schools. The governance of public higher education systems is centralized in many states, which makes it possible for a statewide board, or a state legislature, to mandate changes for an entire system. Since governors and legislators are increasingly concerned about attracting or retaining industry, particularly high technology industry, they can target appropriations for particular programs or alter curriculum requirements through the legislative process or through policy decisions of boards of regents. The governance of the public schools is still comparatively decentralized so that sweeping statewide changes are more difficult to accomplish. Many institutions of higher learning are also in a position to reallocate resources within a school because they rely extensively on part-time teachers who can be hired and fired at will. This is particularly true of community colleges but has become increasingly the case at four-year schools as well. Public schools have heavily tenured staffs with few temporary employees.

Most of the corporate and political concern about high technology work force requirements has focused on the need for more engineers and computer scientists. Government, education, and industry leaders have been saying repeatedly that since the strength of the economy is based on a well-educated pool of labor, it is essential that colleges and universities expand, upgrade, and develop technology-oriented fields of study. There is a fairly clear connection between education and economic productivity when it comes to this level of education. Without the presence of MIT and Stanford University, the Route 128 electronics complex and Silicon

Valley would not exist. These companies' survival depends on the brainpower of the top graduates of schools such as these, and the graduates of lesser-known schools provide critical support as well.

In addition, companies have directed their concern to the training of technicians and have given some attention to comunity college programs. Business leaders have been critical of the private and public colleges for not having programs of sufficient quantity and quality in technical fields. But corporate leaders have muted their criticisms about the response of higher education to their work force needs as student enrollments in the engineering, computer science, and technician training programs swell and as they become aware of the fiscal constraints that limit the ability of the schools to respond to market forces.

Higher education officials are now committed to reorienting their institutions in a more technical direction. However, they lack the financial resources to develop and operate the kinds of programs that companies want. The schools lack equipment, sufficient operating and maintenance budgets, adequate monies to attract new faculty, and necessary scholarship support for graduate students. Despite the extensive publicity given to this problem, no comprehensive and extensive program of support from governmental sources has been forthcoming. Increased corporate donations and other kinds of support from the business community have been helpful, but the financial constraints remain. College administrators, operating within rather severe fiscal limits, have made a considerable effort in the 1980s to build programs that mesh with economic work force requirements.

The nature of the work force requirements themselves has also been a thorny issue for education and government officials. A hot debate exists about whether or not there is actually a shortage of engineers, computer specialists, and technicians of various kinds. Companies and their associations argue that the shortfall of trained personnel is real and continuing while professional societies of engineers and some scholarly analysts insist that increases in the numbers of newly trained employees in those fields is not necessary. With the debate still unresolved, college and university administrators are cautious in their efforts to alter the curricula of their institutions. They are eager to appear responsive to fundamental economic shifts and to satisfy student demand for technical courses, yet they fear that volatile business trends and automation may create unstable job markets for their graduates. Thus, administrators have had to make painful decisions about the reallocation of scarce resources at a time when experts themselves cannot agree on specific occupational projections. By and large, they have proceeded with making changes but have hedged somewhat on the size of the alterations because of uncertainties about the long-run job outlook in technical fields.

Higher Education in California and Massachusetts

California and Massachusetts provide contrasting studies of the ability of states to respond to demands for technically trained workers. While Massachusetts has historically been generous in its commitment to public elementary and secondary schools, its support for public higher education has been niggardly. The reverse is true in California where public higher education has flourished and, in the early 1980s, public schools have been underfunded. The two states also differ dramatically in the percentages of college and university students attending private schools, and they vary as well in the proportion of students attending different levels of higher education.

Enrollment in private institutions of higher education has always been high in Massachusetts. Fifty-nine percent of college students there compared to 22 percent nationwide attend such schools.[1] Massachusetts ranks forty-fourth among the states in its percentage of full-time students who are enrolled in public colleges and universities. California, on the other hand, ranks third among the states in its percentage of students attending public higher education.[2] The two states vary significantly from the national average in their mix of public student enrollments by type of institution. A relatively small percentage of students who are enrolled in public higher education in the Bay State attend community colleges, 25 percent, compared to 36 percent nationally and 52 percent in California. (It should be noted, however, that while California community college enrollments are high, their graduation rates are low. In the six community colleges in Silicon Valley, there were only 2,800 associate's degrees awarded in 1981 out of 90,000 students enrolled.)[3] Community colleges in California cater not only to lower income and working-class students but to many middle-class and upper-middle-class pupils as well. In Massachusetts, these schools have considerably lower status and enroll students primarily from working-class families.

The state college system in California also has much higher status than is the case in Massachusetts. San Jose State University, for example, is a large and well-known institution. Boston State College, by contrast, a small and grossly underfunded school with little political clout, was closed in 1982. Overall, the public higher educational system in California has been exceptionally strong, including an extensive community college system as well as prestigious research universities, while that of Massachusetts has been comparatively weak.

California has spent large sums of money to support its system of public higher education and Massachusetts has spent relatively little. In 1984–85, for example, California ranked seventh among the states (and Washington, D.C.) in appropriations per public college student while Massachusetts ranked thirtieth in its appropriations per student. In 1981–82, Massachusetts ranked fifty-first in the percentage of state tax revenue it allocated to its public colleges and universities while California ranked tenth.[4]

These differences in spending are particularly apparent in examining the condition of the community colleges in the two states. The Massachusetts two-year college system, which has only been developed in the last two decades, has been notoriously underfunded. Faculty salaries, for example, averaged $22,962 in 1982–83 compared to the national average of $25,627 and the California average of $31,715 (second highest in the nation).[5] The fifteen schools have relatively small enrollments and some have shockingly inadequate physical plants. The "campus" of Boston's Roxbury Community College was housed in a condemned nursing home until 1982 and that of Middlesex Community College, located in the center of high technology industry on Route 128, is located in a Veteran's Hospital. They stand in stark contrast to the beautiful and well-funded community colleges in and around Silicon Valley. Approximately 7,000 day and part-time evening students are served by Boston's two community colleges compared to 20,000 students in San Jose, a city roughly the same size as Boston.

The community colleges have played a major role in California's economic development both in training new workers for industry and in retraining older employees. But because community colleges have been so small and ill-equipped in Massachusetts, profit-making proprietary schools and other education and training institutions have played a more important role in vocational education. For example, Sylvania Technical Institute, a proprietary school, has long been the major source of newly trained electronics technicians in the Boston area whereas the bulk of such training in Silicon Valley is done in public community colleges. Until the fall of 1984, the California community colleges charged no tuition and, even now, it is only $50 a semester.

Community college administrators in Massachusetts have learned to run courses on a shoestring. Lacking money from their regular budgets, they often offer courses on a revenue-generating continuing education basis. And most of the money used to start up new technology-oriented programs in the last few years has come from federal occupational education funds administered by the State Department of Education. These funds, however, diminished in the early 1980s as a result of federal budget cutbacks. Several of the colleges in the Boston area have had to turn away students because of inadequate space or lack of permanent faculty.

The fiscal situation in the two states changed somewhat in the early 1980s with Massachusetts providing increased support for its system and California providing less. Between the 1981–82 and the 1983–84 academic year, appropriations for public higher education in Massachusetts went up 24 percent, the second-highest increase among the states. In California, during that same period, appropriations actually declined by 5 percent (a 14-percent decrease when inflation was taken into account).[6] Buoyed by a prosperous regional economy, Massachusetts continued in 1984–85 to appropriate relatively high real increases for colleges and universities. It was California college educators who were protesting inadequate budgets in the early 1980s, a condition that has been chronic in Massachusetts. During 1983–84, the 106 California community colleges sustained a 7.2 percent cut in their budgets (before inflation) resulting in the layoffs of 10,000 hourly instructors and other staff, the loss of 160,000 students and the elimination of 15,000 classes. Twelve community college districts feared the possibility of bankruptcy during that year. Faculty pay at the University of California and state colleges began to lag behind that of comparable schools, buildings and equipment were not being repaired, and new materials and equipment were in short supply.

However, by the mid-1980s, the California state government,

benefiting from the increased revenues of an economic upturn, began to pump money once again back into public higher education. The substantial sums allocated to the University of California and state university system in 1984–85 represented a reversal in the recent financial deterioration of those schools, and the imposition of tuition at the community colleges promised greater revenues for those schools.

An interesting phenomenon occurred in the two states during the first half of the 1980s: They weakened their traditionally strong areas of education (public K–12 education in Massachusetts and public higher education in California), while they strengthened their weaker areas (public colleges and universities in Massachusetts and public K–12 schools in California).

High Technology Programs:
The Response of the Community Colleges

Community colleges are the segment of the educational system that is most responsive to emerging corporate needs. The colleges are mandated to respond to local market forces, teach increasingly older students who are aware of career opportunities in high technology fields, and have financial incentives that propel them to structure courses to fit industrial trends. National data indicate that two-year post-secondary degrees awarded in high technology fields grew modestly during the 1970s and then grew dramatically after 1979. A 1984 study by W. Norton Grubb shows that by 1982, about 19 percent of all associate degrees (and other pre-bachelor's awards) earned were in high technology areas of study, up from 13 percent in 1971.[7] It is apparent both in the Boston area and in Silicon Valley that community colleges are in a position to start and end courses of study quickly, cater to a variety of student populations, run courses of variable length, and provide companies with customized training for their employees. The growing vocational emphasis of their curricula makes them philosophically inclined to alter programs in congruence with labor market trends.

The function of community colleges is still being debated with many faculty and policymakers arguing that the schools should retain a liberal arts emphasis that will allow students to transfer to four-year colleges. But it is clear that a variety of forces, including student demand, have pushed the two-year colleges to allocate greater resources to career-oriented programs.

COMMUNITY COLLEGES IN SILICON VALLEY

The community colleges in Silicon Valley have a greater capacity to respond to new industrial trends than do the Boston area public two-year schools. Although they experienced cutbacks in the early 1980s, the six colleges in northern Santa Clara County had relatively generous funding up to that point, whereas the Boston area schools have never enjoyed adequate financial support. The variety and size of programs relevant to high technology companies in these six schools is impressive. Mission College was established in 1979 with a mandate to coordinate programs with the neighboring electronics complex. DeAnza College in Cupertino has been aggressive, sophisticated, and successful in establishing training links with high technology firms, and Foothill College in Los Altos has a reputation for high quality technology programs. Its A.S. degree in semiconductor processing was the first of its kind in the nation, and it started a new computer science program in 1983 after Tandem Computer gave the school a million dollars worth of equipment.

San Jose City College has the oldest electronics technology program in the Valley and also offers a range of other courses relevant to electronics firms. The offerings at West Valley and Evergreen Community Colleges also include a number of courses and degree programs in technical areas. Enrollments in electronics technology and computer programs have increased significantly in the late 1970s and 1980s. "We could fill up all the community colleges in the Valley with just electronics students" claimed one administrator. Several of the colleges teach approximately 1,000 students each year in either their electronics or computer courses. Foothill can now offer computer classes to more than 5,000 students and plans to require computer competency of all its graduates. All of the community colleges in the Valley report that electronics and data processing, along with business, are their most popular courses. Still, these schools are constrained by financial factors from fully meeting student and industrial demands for technically trained workers. There are often long waiting lists for admission to these classes. Hundreds of students are turned away (one administrator estimated that thousands are turned away in a given year in the two schools in his district) because of lack of equipment and other resources needed to teach them in these subjects.

Further, the rapidity of technological change in the electronics industry poses special problems for the community colleges. "Everything is in a change mode" observed one educator. Like two-year post-secondary schools around the country, these institutions frequently lack state-of-the-art equipment, have a shortage of trained full-time faculty, and have difficulty keeping veteran faculty members abreast of new developments in their fields.[8] All of the community college personnel interviewed in Silicon Valley felt that it was extremely difficult if not impossible for them to have good equipment for educational purposes. Some equipment is simply too expensive to buy. If schools are to have successful training programs in certain fields, they will have to rely on the use of a company's facilities, a phenomenon that occurs infrequently. (Companies are concerned about equipment breakage. High technology firms are especially sensitive about protecting production secrets.) The fields of computer-assisted drafting, microwave technology, and vacuum technology are examples of programs where the utilitization of company equipment on-site is necessary. And even when new equipment is obtained, it rapidly becomes outdated.

Faculty become obsolete as well in their areas of expertise. One community college administrator spoke despairingly of a full-time electronics instructor "who sees computer-assisted design as a fad." Some schools appear to have been more successful than others in utilizing incentives to insure faculty competence. One administrator successfully forced faculty to retrain in their fields by threatening to lower their pay.

Recruitment of new faculty in technical fields is also difficult because of competition from higher paying industrial jobs. Some colleges rely heavily on part-time teachers who are "moonlighting" from their positions in high technology firms. The engineering and technology division of one college, for example, employs 10 full-time staff and 90 to 120 part-time teachers. There are obvious problems associated with such massive temporary hiring: difficulty in evaluating teaching performance, constant hiring efforts, and lack of regular faculty to develop curricula and handle department business. However, the advantages are that the teachers are up-to-date on industrial practices and curricular emphases can be shifted quickly without having to retrain tenured faculty. The part-time instructors often provide advice on updating the curriculum. At the same time, there are only a handful of qualified applicants for full-time positions in technical departments who can teach courses during the day. An opening for a professor in semiconductor processing at one school did not have a single applicant for an entire year. Because of this general problem, some courses can only be taught in the evening when part-time faculty are available.

The Silicon Valley community colleges suffered during the first half of the 1980s as a result of the general cutback in funding for higher education in California. In the words of an administrator of one community college district, "We've been murdered by budget cuts. There is almost no money now for new equipment. In the early 70s we had $600,000 to $800,000 for capital outlay for the two campuses in the district. Since Proposition 13, we have no more than $100,000 a year." The two colleges in San Jose sustained a budget cut of $3.3 million for the 1983–84 academic year, the West Valley district colleges lost $1 million, and the Foothill-DeAnza district had a budget reduction of $2.8 million. Of the six schools in Silicon Valley, five experienced enrollment reductions of from 6 to 11 percent in the fall of 1983 after courses were cut and fears of a tuition charge created uncertainty among students. San Jose Community College alone had 1,465 fewer students than the previous semester as a result of a large cutback in course offerings, particularly at off-campus sites. ("Frills courses" that were avocational or recreational or were for personal development were eliminated from the tuition-free catalog in previous years.)

The community colleges had to dip into their reserves to balance their budgets for the 1983–84 school year. Many part-time instructors were laid off, including half of those in the Foothill-DeAnza district. As a result, many evening courses were canceled, including some in electronics and other technology-related fields. Even though some community college funding cuts were restored during 1984 and 1985, the colleges still felt they were on shaky financial footing. For example, the budget restorations in 1984 only returned the colleges to their 1982–83 spending levels. The additional money still left some community colleges (such as the San Jose Community College district) with deficits, especially since locally imposed laboratory and student health fees, among others, were abolished by the state when tuition was mandated. It takes years for schools to rebuild programs that have been retrenched. And uncertainties in funding have continued to disrupt efforts to plan the curriculum rationally. According to Richard Goff, superintendent of the San Jose Community college district, "We are incapable of planning. We have had six different finance plans come out of Sacramento in the past eight years. . . . Our long range planning does not seem to extend beyond next Monday."[9] An aggressive attempt by the Foothill-DeAnza district in 1984 to have voters approve a special tax to boost the schools' revenues for equipment, books, and maintenance failed to receive the necessary two-thirds support of the local electorate.

Still, it is important to keep in mind that these schools have generally well-regarded programs and facilities compared to community colleges in many other states. The fact that their recent financial plight has received so much press attention shows how important these institutions are to Californians. By contrast, the Boston-area papers rarely report on the more severe condition of the local community colleges. Several of the Silicon Valley community college districts also had a cushion of financial reserves to fall back on during the lean years, and two districts profited from one-time monetary windfalls when they sold valuable property in the mid-1980s. Funds from these sales shored up equipment and maintenance budgets.

THE BOSTON AREA COMMUNITY COLLEGES

Public community colleges in Massachusetts are small and poorly funded. But despite their lack of resources, some of the schools have developed a range of technology-oriented offerings and student enrollments in them are growing. The fifteen community colleges in the state more than doubled their enrollments and degrees awarded in technology programs (mostly in computer programming, data processing, electronic technology and engineering transfer programs) between 1976 and 1982. The disbursement of federal occupational education funds administered by the state to secondary and post-secondary schools shifted in the direction of funding programs in high technology areas after the late 1970s, a shift that was critical to the development of new and expanding programs.

All of the Boston area community colleges have increased enrollments in electronics courses and even greater increases in computer-oriented programs. Several schools, particularly Northern Essex, Bunker Hill, and Middlesex have made a concerted effort to develop new programs in high technology areas. Middlesex, for example, has courses to train ultrasound technicians, electronics technicians, electromechanical drafters, radiologic technicians, computer operators, and technical writing software personnel. Bunker Hill and Northern Essex offer programs to train technicians and have courses in various data processing technologies. Northern Essex has several hundred night students using the electronics and other technical equipment at a nearby vocational-technical high school, one of the few good examples of a community college degree program that has linked up with the better-equipped facilities of the vocational-technical schools. Northern Essex has become particularly adept at securing external funds from a variety of sources in order to develop new programs.

Massachusetts Bay offers a popular computer science major, training for test technicians, and a word processing program. Roxbury has a program in electromechanical drafting and has a sprinkling of other technical courses. It has plans to link up in a major way with the facilities of the Occupational Resource Center of the Boston public school system for some occupational education programs. Minuteman Regional Vocational Technical High School in Lexington and Central New England College began a Technical Scholars Program in 1984 that allows gifted students to earn a high school diploma and an associate's degree in four years.

Given the fiscal constraints of all of the community colleges and the limits placed on some of them by their inadequate physical plants, it is remarkable that the schools have been able to respond to the market for technical training as well as they have. Some feel, in fact, that they are training beyond their capacity as students line up for a limited amount of time on computer terminals. The effects of chronic underfunding are apparent at all of these schools. Roxbury Community College from its founding in 1973 until 1982 had a facility that was totally unsuited for technical course offerings and the college was forced to offer mostly liberal arts classes. Moreover, the college, which serves a mostly minority student population, did not receive a single penny for capital expenditures during its first decade of operation. Though this scandalous treatment has not been repeated to the same degree elsewhere, other colleges have similar complaints.

All the community colleges claim they have difficulty in hiring technical instructors because the salaries the schools offer are too low. As a result, many rely on hiring retirees. And all report critical understaffing, one result of which is a lack of personnel to develop cooperative programs and to seek donations from industry. One school was relying on two inexperienced volunteers to approach industry. Another school had to turn away 600 students from computer courses in one year because of a shortage of faculty. Some programs, such as microwave technician training, cannot be offered because the equipment is too expensive to purchase, and because companies have not volunteered the use of their own equipment for colleges' training purposes.

The impact of underfunding is dramatically illustrated by the plight of one small, new electronics technician training program at one of the schools studied in 1982. Because there was no money to hire more than two permanent faculty, the course of study could not at that time be expanded to a full two-year degree program. There was no laboratory assistant and there was no money for service contracts for repairing equipment. For a year and a half, the program director, who had many other responsibilities, was without even a telephone for making calls outside the college. Not surprisingly, an employer advisory board for the electronics program had never been established. According to the administrator, "I shouldn't be running this program. I don't know anything about electronics and computers. I can't talk about it intelligently with other people in the field." As the interview was being conducted for this research, his office had to be vacated because of flooding from a rainstorm. The capacity of the California community colleges to have large, high-quality programs far outstrips that of Massachusetts.

Other post-secondary training sources in the Boston area are also expanding programs and course offerings in fields relevant to high technology. Wentworth Institute of Technology, long regarded as a source of well-trained technicians, embarked on a program in the early 1980s to double its enrollments to 5,000 students. Sylvania Technical Institute, a proprietary school, expanded its electronics technician training by one-third in 1979 so that it now trains about 2,000 students a year in its nine-month program. Control Data Institute, a proprietary school preparing students to be computer programmers, computer technicians, and operators, grew by one-third in 1981. Blue Hills Technical Institute, a public two-year institution with 560 students has experienced a significant upsurge of student enrollments in its electronic technician and computer programming associate's degree programs even though its tuition was raised from $500 to $2,000 after the passage of Proposition 2½.

Overall, then, enrollments in technology-oriented programs are soaring at the two-year schools, and the administrators of these institutions are engaged in vigorous efforts to develop a variety of programs and courses relevant to high technology. But budgetary constraints at the public community colleges are choking the flow of newly trained personnel from those schools to industry.

Engineering and Computer Science at Four-Year Colleges and Universities

While high technology industry is increasingly worried about the responsiveness of public schools and community colleges to their needs, they are most concerned about the shortage of engineers, particularly electrical engineers, and computer science graduates from undergraduate and post-graduate degree programs. These are the employees who can make an immediate difference to a company. The American Electronics Association (AEA), based in Palo Alto, and the Massachusetts High Technology Council (MHTC) have widely publicized their own surveys, which project a shortfall of engineers and computer scientists. The most recent AEA projections covering the period 1983–87 predict a nationwide shortfall of 113,406 computer science and electrical engineering graduates.[10] A 1982 survey of member companies published by the MHTC projects an annual demand for at least 1,765 new electrical engineering and computer science graduates through 1985 at a time when a pool of only 945 graduates would actually be available for hire each year from Massachusetts schools. (The survey assumes that 30 percent of the graduates would go out of state for work and the bottom 10 percent of the classes would not be qualified for employment at the companies.) And even if all of the new graduates in the state were hired each year, there would still be a shortfall of these technical professionals.[11] State sponsored work force projections for Santa Clara County and for Massachusetts (where most high technology work is done in the Boston area) show continued strong demand for engineers and computer scientists.[12]

The message conveyed by the mass media that well-paying engineering and computer science jobs are plentiful has not been lost on students. Attentive to the job market, college students across the country have been flocking to majors in these fields. Engineering is now the most popular intended area of study among male high school seniors taking the College Board Scholastic Aptitude Test. (Table 6.1) A little over 21 percent nationally chose engineering in 1984, with similar percentages doing so in California and Massachusetts. Less than 4 percent of females, however, indicated the same. Interest in engineering rose every year between 1975 and 1982, declining slightly after that. Growth in interest in computer science has been phenomenal. The proportion of students among SAT test takers choosing computer science has grown sixfold since 1975 and grew by one-third between 1982 and 1983 alone. Ten percent of students now say they want to major in that discipline.[13] Undergraduate engineering enrollments, which have historically gone through periods of boom or bust, have been on the upswing since a trough in the mid-1970s. (Figure 6.1) Entering freshman classes reached an all-time high in 1982 and would have been higher if colleges and universities had been able to accommodate the demand. Many schools have had to resort to mechanisms to restrict the number of students entering these fields, which has had the effect of "weeding out" all but the brightest pupils. Graduate enrollments are up as well.[14]

These national trends are apparent in both the Boston area and Santa Clara County schools of engineering and computer science. The New England and California schools already educate a dispro-

TABLE 6.1   Percentage of 1984 College Seniors
Taking SAT Exams Choosing Engineering or Computer Science
as Their Intended Area of College Study, by State and Sex

	ENGINEERING*		COMPUTER SCIENCE/ SYSTEMS ANALYSIS†	
	Male	*Female*	*Male*	*Female*
National	21.4%	3.6%	12.1%	7.7%
Massachusetts	20.4	2.9	11.6	5.9
California	21.5	3.9	11.5	6.8

*SAT   verbal mean: 453 (national)
            math mean: 543
†SAT   verbal mean: 411 (national)
            math mean: 483

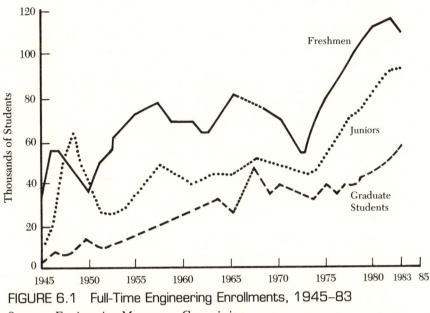

FIGURE 6.1　Full-Time Engineering Enrollments, 1945–83
SOURCE: Engineering Manpower Commission.

portionate share of the nation's engineers. California educates more engineers at all degree levels than any other state, and Massachusetts ranks sixth among the states in awarding undergraduate engineering degrees, and is fourth in both master's and doctoral degrees. Stanford University and MIT are the top two schools in the nation in the production of master's and doctoral degrees in engineering. The University of California at Berkeley, close to Silicon Valley, is ranked fourth when it comes to master's degrees and is third in the production of doctoral degrees.[15]

The number of bachelor's degrees in engineering awarded at the three Massachusetts public universities jumped by 53 percent between 1977–78 and 1982–83 and increased by 57 percent at the state's ten private colleges and universities with engineering programs. (Private institutions of higher education in Massachusetts account for two-thirds of all engineering degrees awarded each year.) The supply of new engineers with baccalaureate degrees has leveled off in the state since 1983–84, however, since almost all of the schools reached full capacity at that time. Growth in electrical engineering and computer science has been especially dramatic with the number of undergraduate degrees awarded in the state doubling between

1979 and 1985. Nearly one-third of all undergraduates at MIT who have declared a major are in electrical engineering and computer science. Master's degrees granted between 1977–78 and 1982–83 grew by 42 percent at the public universities and by 34 percent at the private schools. But the number of doctoral graduates during that same period increased only slightly at the University of Massachusetts (from twenty-four to twenty-six graduates) and actually declined at MIT (from 177 degrees granted to 171), the source of almost all the privately educated doctoral recipients in engineering in the state.[16]

The University of Massachusetts at Boston has expanded its engineering transfer program so that after two years, students can transfer to one of five other colleges (including four private schools) to receive an engineering degree while continuing to pay the lower public tuition fee. The university also started a master's degree program in biotechnology and biomedical science to meet the growing needs of biotechnology firms in the Boston area. Over 1,000 students are enrolled as computer science majors in the nine state colleges.

Boston University has begun a major reorientation of its undergraduate curriculum in the direction of science and engineering. The number of undergraduate students enrolled in engineering and computer science has soared in recent years, and it is the only Massachusetts engineering school still allowing its enrollments to expand. Advanced degree programs in business administration aimed at technical professionals have been introduced as well. The number of undergraduate engineering degrees granted by Worcester Polytechnic Institute (WPI) jumped from 300 in 1977–78 to more than 400 in 1982–83. WPI, along with Bentley College, Boston College, and MIT, is one of the top four schools in Massachusetts in producing bachelor's degree recipients in computer science.[17] Lesley College in Cambridge led the way in developing graduate programs in the use of computers in schools. Central New England College, a technology-oriented school, opened a large satellite campus in 1984 in the Westborough Business Park, which houses leading high technology firms. Business representatives have played an important role in shaping the college's curriculum.

Northeastern University, the largest supplier of technical professionals to New England high technology firms, established a separate College of Computer Science in 1982 whose enrollment by 1984 had almost quadrupled. Enrollments in engineering increased

30 percent in the early 1980s, and a dual engineering degree program was established with neighboring Emmanuel College. Northeastern has also developed a range of other new programs to respond to growth in high technology employment: a part-time High Technology Master's in Business Administration aimed at experienced technical professionals; various kinds of computer technology offerings; technical writing programs; and special career transition programs for women in engineering and information science.

Similar trends have occurred in California schools. Total undergraduate engineering degrees increased 33 percent and graduate degrees rose 9 percent between 1978 and 1981. Even greater increases were recorded in engineering fields most relevent to high technology—electrical, aerospace, bioengineering, and computer engineering—which rose 44 percent during that period, and in computer science, which soared by 68 percent at the undergraduate level between 1978 and 1980, a trend that continued into the 1980s.[18] Enrollments in undergraduate engineering at San Jose State University, which supplies more engineers with bachelor's degrees to Silicon Valley firms than any other school, have more than doubled in recent years. At nearby University of Santa Clara, a Jesuit institution, majors in engineering jumped by one-third in the late 1970s and early 1980s. Stanford University is increasing its output of systems-oriented scientists and engineers in master's and doctoral degree programs affiliated with its new Center for Integrated Systems, a center for microelectronics research that focuses on the development of very large scale integrated systems (VLSI). The Center is projected to produce thirty Ph.D.s and a hundred master's degree graduates each year.

These trends are only a sampling of the alterations in curricula and enrollment shifts into technically oriented programs that are occurring in public and private colleges. From all this it is clear that colleges and universities are responding to changes in the regional economic environment, although industry may not see these changes as being as rapid or pervasive as they would like. The changes appear to occur not because of the direct influence of high technology employers on university policies but because of students' demands for technologically oriented fields of study. In an era of declining enrollments, colleges have to be responsive to their own market, their students. When students clearly want access to certain kinds of programs, the schools feel they must respond to some extent if for no other reason than their own survival as viable institutions. Major curriculum shifts in higher education are not usually initated by faculty and administrators prior to a change in student interest.

Obstacles to Expansion of Engineering and Computer Science Programs

Engineering schools and computer science programs across the country have not been able to expand sufficiently to cope with the rising student demand. The most serious problem is the difficulty of attracting faculty because of industry's superior research opportunities and higher pay. Approximately 10 percent of engineering faculty positions are vacant (between 1,400 and 2,000 vacancies annually) and a 1983 study reports a 17 percent vacancy rate in computer science and computer engineering departments.[19] Fiscal constraints of the schools, due in part to declining federal support, have led to a deterioration of research and teaching facilities. Teaching equipment at most universities is twenty to thirty years old and most engineering laboratories and classrooms are thirty years old.

Class sizes are now larger because of increased student enrollments and the large number of faculty vacancies. Research grants are more difficult to obtain. A large majority of engineering schools report that the quality of both research and teaching has declined. There is a greater reliance on teaching assistants and adjunct faculty to teach courses and many courses cannot be offered at all.[20] A 1983 national survey of faculty members in all disciplines found that 70 percent of the professors in engineering and computer science thought their departments were not able to fulfill their mission, the highest percentage saying so of any of the disciplines studied.[21]

The number of engineers and computer scientists with newly acquired doctorates, the traditional source of new faculty, has not been increasing. There were fewer engineering doctoral degrees granted in 1984 than there were ten years earlier, although enrollments of full-time graduate students in doctoral programs are now at an all-time high after declining significantly in the mid-1970s. (Figures 6.1 and 6.2) There has been a large decline of doctoral degrees granted in the field of electrical engineering: In 1972 there were 850 such degrees awarded, but that number had shrunk to 503 by 1981. After 1981, a modest reversal began, so that

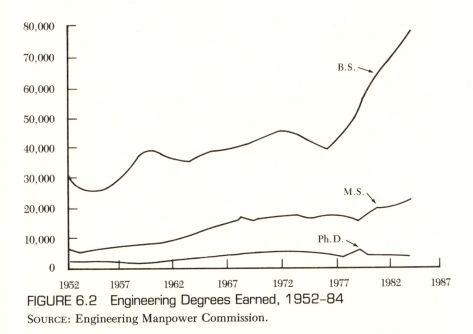

FIGURE 6.2   Engineering Degrees Earned, 1952–84
SOURCE: Engineering Manpower Commission.

by 1984, 693 degrees were granted. The number of doctoral degrees granted in computer engineering also decreased between 1979 and 1983.[22] Only around 250 people annually receive doctoral degrees in either computer engineering or computer science.[23] Further, 40 percent of those who received doctoral degrees in engineering in 1984 were foreign nationals, at least a third of whom do not remain in the United States.[24] However, almost one-fourth of all junior faculty now in engineering departments received their undergraduate training outside the U.S.[25]

There are several reasons for the shortage of doctoral candidates and new faculty: high industry salaries prompt some faculty to leave academe and discourage younger students from pursuing teaching careers; class sizes are substantially larger than before; research support is more difficult to obtain; and university facilities for research are outmoded compared with the more exciting state-of-the-art research operations available in industry.

It is the difficulty of attracting new junior faculty rather than the retention of senior faculty that presents the major problem.[26] A National Science Foundation survey of graduate engineering departments showed that 6 percent of the engineering faculty in these 1,500 departments resigned during 1978–79, and only 28 percent of those left in order to take a nonacademic job.[27] A 1983 survey conducted for the *Chronicle of Higher Education* found that current engineering faculty were only a little more likely than faculty in all other disciplines to be seriously considering leaving for a job outside of academe (15.6 percent compared to 11.5 percent in all fields). Yet that same survey showed that engineering departments were significantly more likely than others to have difficulty recruiting qualified full-time faculty.[28] Faculty in engineering and computer science average extra earnings each year equal to 38 percent of their academic salaries. While computer science and engineering salaries have typically been more than $3,000 higher than those of professors in other disciplines and though they have recently been receiving pay increases greater than that of their colleagues, there has still been pressure in many schools and states to place engineering and computer science faculty on a pay scale even further above that of other faculty. By 1983–84, full professors of engineering in public institutions received almost $10,000 more than their colleagues in other fields.[29] The difficulty of attracting adequate numbers of new faculty and the lack of other resources mean that only one in three qualified engineering undergraduate applicants are accepted into an engineering program.[30]

The Silicon Valley and Boston area engineering and computer science schools have experienced all of these difficulties although the more elite schools, notably MIT and Stanford, have suffered less than others. But even MIT has had trouble attracting the faculty it needs in some fields. The engineering programs at the publicly supported universities have been especially hard hit. A 1984 study commissioned by the Massachusetts Board of Regents of Higher Education reported that $13 million was needed immediately to purchase and upgrade equipment for the engineering schools. The team of outside evaluators noted that the University of Massachusetts had only $50,000 a year to purchase capital equipment, a figure they labeled "completely inadequate." They cited the fact that the University of Lowell has only one secretary for thirty faculty members in the electrical engineering department. And they pointed out that Southeastern Massachusetts University could not install a major computer and 100 terminals for more than half a year because there were no technicians to install the equipment.[31]

In 1982–83, there were 633 applicants to the computer science program at the University of Massachusetts at Amherst, but there were facilities to handle only the twenty-one who were accepted. Their average SAT scores were 700 in mathematics and 650 verbal, about 200 points higher than the average of other freshmen.[32] In 1982, there was a shortage of 183 engineering faculty at the three Massachusetts state universities, including thirteen at Lowell in electrical engineering alone just to handle students who were already enrolled there. Overcrowded classrooms remain a major obstacle to the improvement of program quality.

Northeastern University has ten faculty vacancies they have not been able to fill. In 1982, it forecast a need for an additonal $3.5 million for equipment purchases and over $1 million more for faculty salary compensation. Salem State College, located near the northern end of Route 128, had turned away as many as 200 students a semester from computer science courses because of faculty shortages. But after hiring three new faculty members, student enrollments in that field rose from 460 to 730 in just one year. The well-regarded computer science program at Framingham State College, also located near Route 128 in the western suburbs, had enrollment quadruple between 1974 and 1982 but for a long period had to limp along with only two full-time faculty and twenty-one part-time teachers. In 1982–83, almost 900 students sought to get into the sixty-four slots available in the computer science department.[33]

The implementation of a graduate program in computer science at the University of Massachusetts at Boston was delayed until 1983 due to insufficient equipment. There are inadequate research facilities for faculty there, making it difficult to retain and attract professors. A former administrator at a Massachusetts state college referred to some of the technical programs in the college's extension division as "low quality," noting that faculty were being paid only $900 a course in 1981 ("an insanity"). "There is not enough money in public higher education to respond to high technology," the former official remarked. "We don't have the computers or the faculty and the programs exist under terribly adverse circumstances."

Massachusetts has embarked on a program to improve spending in public higher education and has earmarked a large percentage of the new money to go toward engineering and computer science programs. Thus a critical start on the problem has been made but several more years of sustained levels of higher funding are needed if more than simply "plugging holes" is to occur.

In Silicon Valley the situation at San Jose State University is especially critical. The school has reached the saturation point in student enrollments, and like many other California engineering schools supplemental admissions criteria have been imposed. Some of the engineering programs at San Jose State are turning away two out of every three qualified applicants. Classes are large and there are waiting lists for many courses. Inadequate facilities and an insufficient number of faculty have meant that the school cannot expand to handle student demand. Even when new equipment is donated, the school cannot afford the expensive service contracts and upkeep required.

San Jose State had thirty-five unfilled full-time faculty positions in 1984 out of the ninty-six allocated to the engineering school. Almost half of the classes there are now taught by part-time faculty, putting the school's accreditation in jeopardy and adding extra burdens of counseling and curriculum development on the full-time instructors. A new assistant professor's salary starts at less than $25,000 while newly graduated bachelor's degree recipients receive starting salaries of $26,000 and up if they go into industry. The state university system has not been able to place engineering and computer science faculty on a higher pay scale than other disciplines as the University of California campuses were able to do in 1982. Assistant professors in these fields at the UC campuses are paid 20 percent more than others while associate professors and full professors received hikes of 10 percent and 5 percent respectively.

The situation at San Jose State reflects the problems California public colleges and universities are having statewide. (The public system educates approximately four-fifths of the engineering graduates in that state.) A 1982 report for the California Postsecondary Education Commission reported that the California State University system employed as many part-time faculty as it did full-time faculty in engineering and computer science. At the University of California a similar heavy reliance on part-time faculty was reported as a result of the inability to recruit new professors. Deans and program directors surveyed indicated that classes were too large, students needed more time to complete their programs for graduation as a result of being kept out of closed courses, and heavy faculty advising loads precluded students from needed individual attention. Moreover, the same report noted that the University of California needed $7.5 million annually for several years to replace aging engineering and computer science equipment and another $12 million was needed each year for equipment in new technologies. The equipment problem at the state universities was described as being even "more severe." Besides that, some of the campuses required major capital investments for remodeling and construction of new facilities.[34]

A new financial burden for academe is the cost of developing and maintaining a computer system for campus-wide computer literacy programs. Colleges nationwide are moving very fast to make computers available to students in a wide variety of subjects. A few colleges are requiring some computer expertise of all their graduates. Faculty in many disciplines are enrolled in in-service training courses at their schools to learn how to use the computer in their classes. The demand among students for computer literacy courses is very strong, and most campuses have not been able to provide the resources to handle the demand. Students across the country wait in long lines for their brief turn on the computer. But the cost of providing computer-oriented classes is enormous. According to one estimate, the capital outlay required for an acceptable computer system installed between 1985 and 1990 would be a minimum of $1,000 per student at a liberal arts college and as much as $6,000 per student at a technically oriented school. The annual operating and maintenance costs would be equal to what schools now spend per student on library expenses, about $100 a pupil.[35] So academe finds itself in the position of having to finance the upgrading of existing technical programs as well as coming up with funds to develop new and expensive computer curricula and facilities for campus-wide use.

The four-year colleges and universities are clearly making an effort to develop or expand programs that are congruent with the needs of the "new information society." But those institutions are struggling to find the staff and resources needed to establish and maintain thriving programs. The commitment is there but the capacity to respond to economic changes is lacking.

Engineers and Technicians: The Work Force Shortage Question

A lively debate has been taking place among labor economists, engineering societies, and industry officials about whether a shortage of engineers exists. Labor force projections by industry are criticized as being methodologically flawed and inflated by firms' overly optimistic expectations for sales, profits, and company expansion.[36] Further, government agencies have issued conflicting reports about the long- and short-term employment markets for engineers.[37]

Some academic observers and representatives of engineering groups argue that if the supply of engineers is as constricted as industry people claim, engineering salaries would be rising faster than they are currently, and older engineers would be utilized more effectively. As it is now, they note, high technology companies prefer to hire new college graduates or workers with only a few years of experience rather than retraining or hiring veteran engineers. Indeed, data from a Massachusetts High Technology Council survey shows that member firms expect that less than 10 percent of their new hires of technical professionals will come from the ranks of personnel with more than five years' experience.[38] Added to that is the phenomenon of "salary compression," which afflicts older engineers whose salary increases are smaller than those given to younger engineers. Those disputing claims of shortages argue that industry simply wants to expand the engineering labor supply in order to depress salaries and to have a better selection of personnel. The U.S. Office of Technology Assessment in a recent study provided partial support for this contention:

Entry-level shortages arise in part because employers prefer to hire new engineers with fresh skills at lower pay. This is an easier and perhaps cheaper way of meeting their needs than coupling the experience of mid-career engineers—many of whom find themselves with increasingly obsolescent skills—with well designed continuing education programs.[39]

According to data provided by the Engineering Manpower Commission, engineers nationwide experienced a decrease in purchasing power from the early 1970s through the early 1980s. Slight real increases in average salaries have occurred since then. Engineering salaries vary by region of the country. Of nine regional groupings, Pacific Coast engineers were paid the second-highest salaries in 1984, a median of $40,600, while those in New England ranked third highest with a median salary of $40,150, a large jump from previous rankings.[40] Earlier studies showed that Massachusetts engineers and other high technology workers have typically been paid less than the average compensation of comparable workers in other states, but the tight labor market brought on by the state's booming economy in the mid-1980s has apparently raised salaries relative to other regions.[41]

Engineering societies have argued that the relatively flat real incomes of engineers in recent years are evidence that no true shortage of engineers exists. Otherwise, they claim, salaries would have risen more dramatically. They also point out that actual shortages would have induced companies to utilize experienced engineers more effectively instead of allowing them to become obsolete. High technology companies argue intensely that the shortage of engineers is real and claim that the lack of trained technical workers is already having a direct impact on their ability to get new products on the market. The response of a vice president for human resources at one Route 128 firm, who was interviewed for this study, reflects the industry perspective:

Workforce planning is in its infancy in electronics companies. We hire too many people, then lay them off. We could hire more technicians [to do engineers' work] and must utilize older engineers better by instituting a sequence of training periods into the career life of an engineer. But there *is* a genuine demand for engineers, especially in software engineering. We need engineers to survive. It's ridiculous to accuse companies of wanting schools to pump out engineers in order to lower the price of engineers. Consider the engineering manager [who puts in requests for new personnel]. His job is on the line to develop a particular product in a particular time at a particular cost. He doesn't care about the wage structure—he just wants the personnel to do the job. We have to get to markets fast with new products or we will be blown away.

By almost any measure, engineering shortages will probably exist in some regions and in some subfields. Demand growth is projected by the Engineering Manpower Commission to be strongest in the field of electrical engineering and in the Pacific and New England regions of the country.[42] Within that field, there is especially strong demand for personnel trained in computer-aided-design and computer-aided-manufacturing (CADCAM), microwave, and manufacturing process. Software engineers, specialists in optics, and computer systems analysts are also projected to be in high demand in coming years.[43]

Company officials in Silicon Valley and in the Route 128 area stress their need for electrical and computer engineers, fields where they feel the demand will hold up for some time. Hiring among many firms in both areas picked up dramatically in 1984 as local economies boomed with the end of the recession and the increase in military spending. Yet even as employment opportunities for engineers and computer scientists grew, the Massachusetts Board of Regents of Higher Education issued a report that claimed there would be no shortage of technical professionals in that state throughout the 1980s, including electrical engineers.[44] The study concluded that the in-migration of engineers to the state, the utilization of scientifically and mathematically trained personnel, and the jump in new engineering graduates would more than meet the hiring needs of firms. Yet local industry leaders disputed the report's conclusions and stuck by their more expansive projections.

Business people acknowledged in the interviews conducted for this study that firms are all competing for a small pool of top flight talent, "the kind who will really make a difference to the company." Or, as one human resources executive put it, "My guess is that there are enough engineers, just not enough *good* engineers." Privately, corporate managers will admit that almost all high technology companies do a very poor job of human resource planning, which results in a failure to keep their experienced engineers up-to-date in their fields. (In the most recent AEA work force study, some companies could not participate in the survey because they do not do human resource planning at all.)[45] As long as engineering obsolescence exists on such a massive scale without periodic and planned retraining, there will be continuing spot shortages of the newly trained engineers the companies are seeking.

Amidst the controversy, a sensible middle ground view among educators has emerged. The majority of the college and university administrators interviewed in this study voiced moderate skepticism of industry projections for engineers and computer scientists. Still haunted by the widespread layoffs of engineers in the early 1970s, which left them with undersubscribed engineering programs, they are reluctant to institute enormous expansion in engineering.[46] They are aware that overall demand for engineers is primarily dependent on the overall state of the economy and the level of military spending. University administrators differ among themselves about work force forecasts even within the same institution, but the following comment typifies the views of many:

"It is impossible to say what engineering manpower needs will be. Industry doesn't know any better than we know. They can't forecast political decisions such as military contracts. How can anyone believe anything?" Boston area university administrator

But while academic officials are wary about growth in engineering programs, they generally believe industry claims about the strength of the long-range job outlook for engineers and computer scientists. Paul Gray, president of MIT, cites three factors that will continue to generate demand for technical professionals: the need for energy alternatives in our society; the mushrooming of applications of microprocessors; and future opportunities for applications of genetic engineering. The administrators agree that even if engineering jobs become scarcer in the future (as they did in 1982 and 1983), technically trained people can find other positions (e.g., in marketing) where their backgrounds are useful. They point to the very low unemployment rate among engineers (less than 1 percent in 1983).[47] A common theme expressed is that "if anyone survives in the future, it will be the individual with technical training." Thus, while there is some questioning of future projections of technical professionals, educators are cautiously proceeding with expansion of programs in the belief that new technologies and their applications are the wave of the future. Their general approach is to expand carefully in a way that will not warp the curriculum of their institutions.

Questions have been raised as well about the demand for electronics technicians of various kinds (e.g., manufacturing test technicians, field service technicians, computer maintenance technicians). In Massachusetts, every type of training institution—private, proprietary, CETA, and public colleges—has experienced some difficulty in placing their technician graduates in jobs during the early 1980s. Even as the recession ended in 1983, the growth in employment in high technology firms in that state was not as great for technicians as it was for other personnel. Demand for technicians appears to be stronger in Silicon Valley than in the Boston area. Companies in the San Francisco Bay Council of the AEA project a 19.1 percent annual increase in jobs for electronic technicians between 1983 and 1987 while Massachusetts firms surveyed forecast a 9.4 percent increase each year during that same period, a little below the national figure of 10.3 percent.[48]

The hiring problem for technicians is not severe but enough slack in hiring exists to make school administrators nervous. This is especially true for those Massachusetts institutions that have expanded training significantly or are in the process of doing so. A 1980 study for the Massachusetts Department of Manpower Development also questioned the need for increasing the supply of newly trained technicians.[49]

Executives at several Route 128 companies who were interviewed said they did not anticipate either a short-term or long-term demand for technicians, especially manufacturing test technicians, and cautioned that educational administrators should not expand programs too quickly. They felt that schools should aim for small programs of high quality. A human resources manager with recent experience in two companies pointed out that there has been a dramatic drop in the need for electronics technicians in those firms in the last year. At the same time, other high techology companies claim they have a continuing demand for technicians.

The economic recession in the early 1980s was one reason for the softness in the market for electronics technicians during that period. But changes in technology have altered the work of technicians in some companies. For example, the products of computer companies are increasingly manufactured with "built-in diagnostics" that inform a field service technician what is malfunctioning. Instead of determining what is wrong with that self-identified circuit board, the technician simply replaces the whole board. These "built-in diagnostics" and the practice of "board swapping" has the result of reducing the skill level needed in the technician's job. "Advance maintainability," building products in a way that they can easily be repaired, is becoming more common. Since almost half of all technicians work on repair of some kind, these technological changes have important implications for that occupation. Companies vary in the degree to which they have deskilled technician's work. Even among firms manufacturing the same product, there are differences in how far they have proceeded in these and other technological changes. Hence, employment projections for technicians can vary substantially.

At the same time there are countervailing trends that argue for both greater numbers of technicians and high skill levels among them, and, indeed, the U.S. Bureau of Labor Statistics forecasts strong growth for electrical and electronics technicians through 1995.[50] Applications of computers to various areas of human activity are just beginning to occur, and many of those applications will require technicians to help develop and maintain them. Products are becoming more complicated and a certain percentage of technicians will have to understand the way they operate. There is also growing use of technicians to carry out tasks formerly performed by engineers, thus alleviating the engineering shortage and increasing the demand for technicians.[51] External political forces also make a difference: The large defense contracts of the 1980s, for example, have accelerated the demand for electronics technicians.

Some of the same questions have been raised in the area of entry-level training for computer programmers who are in less than a four-year computer science program. Again, there is widespread agreement among corporate trainers, particularly in the Boston area, that inexperienced programmers trained in fairly short-term programs are having some difficulty finding their first job. When they are hired, it is usually with a "user" firm (a bank, an insurance company) and not with a high technology manufacturer. Entry-level technical writing jobs have not been plentiful either. This mixed picture makes it difficult for educational administrators at various levels of training to make decisions about the allocation of resources to training programs. It is clear that educators need to keep in regular touch with a representative sample of employers, especially in the area of test technician training.

Higher Educational Directions and Economic Change

Higher education officials, then, find themselves in the difficult position of trying to respond to student demands for technical courses while trying at the same time to second-guess labor force trends. "Education is in an uncomfortable mode," observed one university administrator interviewed for this study. "We don't know what to believe about manpower projections but student demand is incredible." While the existence of shortages of new entrants to certain fields remains an open question, the shortage of resources for training in technical subjects is a certainty. The condition of public higher education in California and Massachusetts illustrates variations among states in their capacity to respond to changing industrial directions: the California colleges and universities, a massive well-funded, and high quality system, is trying to maintain that quality in the face of escalating costs; the Massachusetts schools, underfunded and neglected in a region where private education is dominant, are

trying to build a decent and comprehensive set of technical programs. In both cases, the institutions are trying to respond to
economic trends by developing and expanding engineering, computer science, computer literacy courses, and two-year technician
training programs. But inadequate equipment and facilities and a
shortage of instructors hinder their ability to alter and develop programs along the lines consistent with the emerging computer era.
While considerably more flexible and responsive than public secondary schools, higher education shares with them the realities of fiscal
austerity.

7 Corporations and the Campus: Collaboration or Contention?

As colleges and universities struggle to revamp and reorient their curricula in a more technical direction to accommodate economic changes, their direct ties with the business sector have grown. Businessmen, educators, and public officials are now heralding a new era of business-education cooperation. National Science Foundation reports on research links between private corporations and universities speak of "the surge in the volume and variety of [research] interactions" between the two since the late 1970s and claim that "we may be at the threshold of a permanent new state of corporate-academic research relationships."[1] Indeed, the ties in some fields and at certain schools are so close that the public interest may not be well served and free academic inquiry may be constrained. Yet these connections are strong only in select sectors of higher education, and business support overall for colleges and universities remains too modest.

Industry–academic ties appear to be strongest in the fields of engineering and biotechnology, a phenomenon that illustrates the congruence between economic trends and higher educational directions. A 1982 article in *Business Week* claimed that "business and universities are forging a powerful new alliance" in engineering and

high technology "parallel with the 19th century partnership between universities and government in agriculture that made the
U.S. the granary of the world."[2] A range of relationships have been
enhanced or developed including cooperative research relationships
of various sorts, corporate donations of fellowships, equipment and
cash, customized training for corporate employees by academic institutions, campus-based technology industrial parks, and consulting opportunities in high technology companies for individual
faculty.

Direct ties between academe and the business community are
nothing new. Many schools have depended on the fortunes donated
by wealthy industrialists and businessmen have long served on college and university boards of trustees. The chemical industry has
had a long history of close association with academia dating back to
the nineteenth century.[3] Historian David Noble shows that the
whole development of engineering education and the engineering
profession was inextricably intertwined with corporate interests.[4]
Business support of academic research played a "small but significant role" prior to World War II and campus-corporate research,
nourished by government funds, flourished during it.[5] Corporate
funding of university research and development continued after the
war although the major increase in federal support, especially in the
early 1960s, caused the proportionate corporate contribution to fall
during that period. (Figure 7.1) The private sector percentage of
support remained at the level of 3 to 4 percent between 1965 and
1978, but the absolute levels of business support for campus research
and development doubled during that same period even after inflation was taken into account.[6] The late 1970s and 1980s have wit-
nessed an increase in industry-education activity, while federal
funds have leveled off, so that the proportional share of industrially
sponsored research and development on campuses is expected to
grow.

The relationship between high technology firms and campus
engineering, biotechnology, and computer science programs is particularly close, especially when research relationships are considered. At least half of all company supported research in universities takes place in engineering departments, with between 6 and 10
percent of engineering research and development being funded by
industry.[7] The extensive study by Lois Peters and Herbert Fusfeld on
industry-university research connections for the National Science

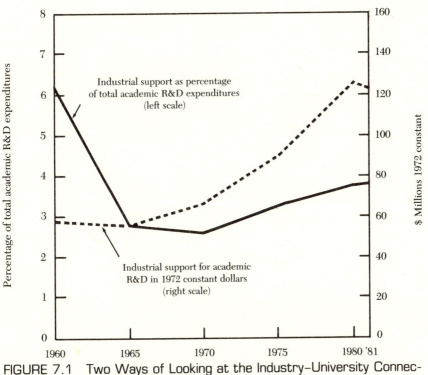

FIGURE 7.1   Two Ways of Looking at the Industry–University Connection

Source: National Science Foundation.

Foundation found that in ten leading research universities, 60 percent of industrially funded research went to engineering, 10 percent to agriculture, and 30 percent to all other technical programs. [8] According to their study, research relations between firms and schools are especially common in interdisciplinary fields undergoing rapid change, such as microelectronics and molecular biology. Generally speaking, large firms are most likely to be engaged in collaborative programs, but in the case of high technology industry, small, innovative companies are sometimes involved in cooperative relations of various kinds with schools as well.[9]

Incentives to Collaboration

There are a number of incentives motivating both corporations and academic institutions to seek out one another for collaborative programs. The needs of the colleges and universities are fairly obvious. As their costs escalate and as traditional sources of support level off or wane, as has been the case with some forms of federal funding, schools look to industry to provide some help with the costs of doing research, constructing facilities, and training both graduate and undergraduate students. Although corporate cash and donations represent a small part of their revenues, the support is still important. Equipment donations can be especially valuable in certain new and rapidly changing fields such as computer science. Higher education administrators welcome business contributions partly because there is less red tape involved than there is with federal financial support.

A second major incentive for the schools to link up with industry is the recruitment of a new population of students. As academe observes the decline in the traditional college-age population, it has aggressively sought out nontraditional students. Many schools, especially community colleges, have become involved in providing training of various kinds for employees in the private sector. And schools of engineering are increasingly offering advanced degree programs tailored to the schedules and needs of practicing engineers. Collaborative research programs with companies offer graduate students the chance to work on interesting problems and open up career opportunities for them. The Peters and Fusfeld study found that university scientists sometimes look to industrially sponsored research as a way to gain access to company research laboratories, which are more sophisticated than their own.[10] Once again, this need is greater for researchers engaged in pioneering investigations in swiftly changing fields.

State governments have also promoted industry-university links in high technology as a way of accelerating training and providing an exciting research environment in order to attract new companies to an area. As a result, a number of public universities are setting up industrial parks on or adjacent to campuses, developing "incubator facilities" to nurture the growth of new technical companies, offering in-service training programs to company employees, and establishing cooperative research programs of various kinds. Thus, economic development concerns explain some of the impetus behind campus-corporate collaboration.

It is commonly assumed that companies support university

research, either in contract form or in gifts or grants, in order to achieve scientific innovations that will lead to the development of new products or processes. But the Peters and Fusfeld research found that the desire for innovative discoveries was not an important reason for why firms sponsor academic research or join industrial affiliate programs. Instead, what companies are seeking first is close contact with graduate students and faculty as a source of trained manpower for the company. Students frequently take jobs with the sponsoring or member firms and faculty become more available as consultants and, sometimes, as future full-time employees.

High technology industry, through the activities of its trade associations and some individual companies, has taken an active stance in response to the engineering education crisis of the 1980s. The American Electronics Association and the Massachusetts High Technology Council have been raising funds from their members to bolster engineering and computer science education and have lobbied for increased government support for these schools. Some large, individual firms, such as Exxon (through the Exxon Education Foundation), are making substantial donations to shore up graduate training in engineering. The trade associations argue frankly that since shortages of technical personnel in certain fields are now critical, and since the quality of training is declining as faculty shortages increase and equipment ages, companies must invest in technical education if they are to have an adequately educated work force.[11] Thus, work force needs explain much corporate activity within academe.

Companies also want a "window on science and technology" when they establish cooperative relations with schools. Access to current knowledge and research activities leads them to become involved in investigative projects of various kinds. This motive, according to the Peters and Fusfeld study, is of special importance to high technology firms:

This ["window"] is a high priority in rapidly changing industries such as genetic engineering and microelectronics, which have recently stemmed from, or have close ties to, the university. For these industries, the technology transfer cycle is very short. They are evolving so rapidly that both the university and the industry must participate in all aspects of the cycle. Frequently, scientists mentioned that access to manpower and obtaining a window on science and technology could not be separated from each other.[12]

Other corporate motives for becoming involved in university research projects include the need to solve a particular problem or to use university research facilities or simply to improve the company's public image.[13]

It is instructive to compare partnerships between companies and higher educational institutions with those involving public schools and private firms. True partnerships are often described as those characterized by genuine reciprocity where both sides have real needs that are satisfied by the relationship. This definition clearly fits many of the industrial-academic connections that currently exist. In many of these arrangements, the universities get needed money or other resources while companies gain access to manpower. Few industry–public school partnerships fulfill genuine immediate needs of companies. At most, firms gain access to a few new employees (whose presence is not critical to company success), improve their civic image, and contribute to the long-range improvement of learning programs. The investment in public education brings rewards in the distant future. The vast majority of these relationships do not really confer reciprocal benefits despite all the rhetoric about partnership that surrounds cooperative efforts. In most instances, the relationship is lopsided with schools needing the help (in donated equipment, for example) from the companies who have greater resources than they and who will derive little in the way of direct rewards for such activity. In other words, most industry–public school programs actually fall under the rubric of philanthropy rather than being categorized as mutually beneficial partnerships. The fact that the relationships are unequal is one reason why they are so precarious.

University officials, especially those in prestigious research institutions, approach industry from the standpoint of greater strength, knowing that they have something the companies truly want. It is no wonder, then, that their relations with the private sector are more harmonious than those of public schools. Less than 5 percent of annual corporate contributions to education (somewhere between $13 million and $22 million) go to pre-college institutions, a dramatic demonstration of corporate priorities when it comes to education.[14]

That is not to say that the industry–higher education connection is a smooth one. This history of the relationship is one of a "pattern of attraction and avoidance" with a good deal of wariness on both sides about cooperative programs.[15] As business contributions to research and teaching become more pronounced, questions about academic freedom and institutional and professorial independence are raised. The issue of ownership of intellectual property rights in collaborative research efforts, while not an overwhelming barrier, has been an impediment to joint work. More importantly, the publicity given to the new alliances between companies and higher education usually overlooks the fact that most of the activity is taking place between a few large firms and the most prestigious research universities. And even there, industrial support makes up only a small fraction of university budgets. Four-year colleges, lesser known universities and community colleges have very fragmentary contacts with companies. In addition, though some high technology firms are at the forefront of developing cooperative programs with higher education, the industry as a whole has a reputation for being relatively stingy with its resources. The much-ballyhooed references in the press and in political speeches to an emerging era in industry–higher education connections are thus often overstated. These ties are still fairly tenuous and unpredictable.

The Route 128 area and Silicon Valley provide interesting case studies of the condition of the relationship between higher education and the high technology companies that depend on well-educated labor. Much of the material for these case studies was gathered through personal interviews conducted with industry officials and higher education administrators in these two regions.

The Community Colleges: Relationships with High Technology Firms

Budget cuts and the prospect of declining enrollments have propelled community colleges across the country into a closer relationship with industry. This trend dovetails with the community service orientation and increasingly vocational direction taken by these schools. While both the Boston area and Silicon Valley community colleges have developed a wider range of collaborative programs with industry, the phenomenon is considerably more pronounced in the California two-year schools. Many community college officials there are fearful of declining enrollment in the future as they see school attendance rates plummeting in their feeder districts. The budget cuts brought about by Proposition 13 have also led schools to seek out industry donations of equipment. A number of industry officials pointed out that as a result of these forces, community colleges have become much more aggressive in seeking to develop cooperative relationships with industry. "Proposition 13 has actually aided the relationship" between the two institutions observed one electronics industry manager in the Valley.

The Massachusetts community colleges, small and underfunded to begin with, have always been funded from the state budget (unlike the California community colleges, which had relied heavily on the local property tax as a source of revenue prior to 1978), and thus were less directly affected by Proposition 2½. Nevertheless, they too have reached out to companies for donations of equipment, work placements, curriculum advisory help, and involvement in corporate training programs.

The variations in the rapprochement between industry and the community colleges between California and Massachusetts illustrate how practices differ from state to state. Some states such as South and North Carolina and Iowa have mobilized these schools in a comprehensive way to meet the training needs of industry, especially high technology industry.[16] The South Carolina Technical Education System was set up in the 1960s to develop crash programs to train workers for companies coming to the state and to continually train large numbers of technicians. In 1984, North Carolina spent $6 million training workers for specific new high technology jobs moving into the state. Others, New Hampshire for example, have set up a coordinating agency to match company training demands with programs at both public and private community and four-year colleges. Firms outline their training needs and the schools then offer specific skills courses of varying length to accommodate that need. Governors have developed these strategies as a way of attracting new companies to their states. More than two-thirds of the states have now set up training programs for specific industries. Companies praise this fast and focused approach, but some states have chosen to avoid such heavy involvement in narrow company-specific training. Neither California nor Massachusetts has developed two-year skills centers or reoriented its community colleges in an organized way to respond to industrial training imperatives. Instead, individual schools go their own way in developing such relationships, with some preferring to emphasize a more traditional comprehensive curriculum, including the liberal arts, and others placing a good deal of time and resources into developing programs that will respond to company needs.

PARTNERSHIPS WITH SILICON VALLEY COMMUNITY COLLEGES

There are numerous examples of partnership efforts between high technology firms and community colleges in Silicon Valley and in the Route 128 area around Boston. The ties are more significant in the Santa Clara Valley schools than they are in the Boston area. Most industry and education officials in the Valley interviewed for this study believe that community colleges are responding reasonably well to the personnel requirements of the firms in that locale. Business executives complain that the colleges do not turn out sufficient numbers of trained personnel for the electronics industry, but most observed that financial constraints make it difficult for the colleges to do much more than they are.

There are a variety of direct and indirect ties between local companies and the community colleges. Employer advisory committees exist in all of the vocational programs, a relationship that has been a fairly satisfactory one from the point of view of both industry and education people. One college went beyond that and hired consultants from industry to develop an electronics curriculum when it was in its formative stages. Another link between industry and the community colleges is provided by the large numbers of part-time teachers who are "moonlighting" from their jobs in high technology firms. Sometimes these instructors will become involved in curriculum development in the schools. Some of the electronics firms have cooperative or work-experience programs involving community college students.

A number of companies make equipment donations to the schools although the gifts are fairly sporadic and cannot be counted on by administrators. Some of these gifts have been fairly substantial. For example, Intel donated $85,000 worth of new equipment to the electronics technology program at Mission College when it was founded in 1979 and has provided it with a steady stream of supplies and materials since then. Hewlett-Packard gave the equipment for a self-paced electronics learning laboratory to nearby College of San Mateo. Intel, National Semiconductor, and Fairchild provided some of the electronics laboratory facilities at Foothill College.

Tandem Computers made a major gift to Foothill in 1983 when it donated more than $1 million worth of computer equipment and services to the school, a gift that allowed the school to triple the number of students it could handle in computer courses. The "Non-Stop" system, which operates twenty-four hours a day, seven days a week, is the first on an American college campus to be used solely for instructional purposes. ("It seems like a good partnership," according to a Tandem executive. "They wanted the system, and we're based here, so we'll need more employees who know our computers.")[17] Computervision, headquartered in Massachusetts, donated approximately $1 million in computer-aided-design (CAD) equipment to DeAnza College, hoping that it would facilitate the training of technicians in that specialty. IBM donated a CAD unit worth $150,000 to the San Jose Community College district in 1984. And a few firms allow the colleges to come in and use their equipment in a course. For example, DeAnza College has used the Kalma facilities to teach a course in computer-assisted drafting.

Community colleges are becoming more and more involved in various kinds of training programs with corporations. For example, Intel and neighboring Mission College have had a program whereby fifty Intel employees receive released-time from the company to attend the electronics technology courses at Mission College until they receive a two-year degree. Then, the students are promoted by the company and return to their full-time employee status. The community colleges have attempted to reach more students by providing courses at sites away from the campus, including company plants. Most of the companies studied offered such courses in their buildings, usually after hours, for the convenience of their employees. Sometimes the courses are part of the firm's training program. In this case, the classes are either funded by the community college as long as they are open to the public or, if they are closed to company outsiders, the company pays for the course (a "contract course"). Occasionally, the firm has some say in the selection of the instructor. The colleges are eager to help provide company training in these ways.

The training needs of the corporations have been expanding in recent years for several reasons: the technology is changing so fast that workers must be frequently retrained; with high housing costs making recruitment from other regions difficult, it makes sense to upgrade and retain workers already on the payroll; and companies need to make up for writing and mathematical deficiencies of their workers. Industry managers interviewed were evenly divided on the issue of whether community colleges should play a greater role in company training programs. According to most observers, the need for community college involvement in corporate training is greatest among small firms who cannot afford an elaborate internal training program, but there are many bureaucratic obstacles to forming training consortia among these companies.

For the most part, business managers were pleased with the quality of community college courses offered in their plants although there were a couple of cases where firms felt that a school's training materials were disappointing. The importance of the Silicon Valley community colleges to workers trying to upgrade their positions in electronics was documented in a 1983 study by Judith Larsen and Carol Gill. They interviewed women working in electronics firms and found that the "community college system was instrumental for career advancement."[18] Most of the women studied reported taking community college courses to aid them in their jobs, especially those making more than $26,000.

The Foothill-DeAnza Community College District is exploring a more elaborate training partnership with local high technology firms. The district is considering plans for a nine-acre, $60-million complex on the DeAnza campus that would include a high technology institute, a hotel, and a conference center. The complex would be revenue-producing by training workers for companies, serving as a site for conferences, and providing accommodations for workers from out of town who come to attend training courses. DeAnza already is heavily involved in corporate training programs (with over $1 million in business training contracts in 1983–84), but such a center would increase its provision of training substantially. If such a complex were developed, it would be the most elaborate high technology training center affiliated with a community college in the country.

Some community college–electronics industry training occurred under the auspices of the California Worksite Education and Training Act (CWETA), which was passed by the state legislature in

1979. The $25-million bill, which was extended through 1984 with an additional $10-million appropriation and then terminated, had its origins in electronics industry leaders' expressions of concern to Governor Jerry Brown that community colleges were doing an inadequate job in training skilled workers. The legislation provided money for community colleges to work with firms in training structurally unemployed workers for entry-level jobs in industry and for training lower level workers for more highly skilled positions. The companies were to provide facilities, released-time, and, finally, a promise of jobs at the end of the training period. The California Employment Development Department (EDD) was given the job of administering the program.

Almost everyone interviewed in this study felt that CWETA was not only "too little too late," but was also so bedeviled by bureaucratic red tape that the program was not worth the expenditure of so much time and money. Industry people criticized aspects of the curriculum of the community colleges and their rigidity in credentialing appropriate instructors, and college people felt that company-based instruction was too narrow. Many companies, caught in an industry slowdown, lost interest in participating when they realized they had to guarantee jobs to those completing the program. Both industry and education personnel criticized the coordination efforts of EDD. An exception to this general reaction in the Silicon Valley area was the apparent success of the $2.5-million CWETA electronics program sponsored by the College of San Mateo.

The functions of CWETA have been replaced by the federally funded Job Training Partnership Act and the state supported $55-million-a-year Employment Training Panel (ETP), a program with more ample and stable funding than that of CWETA. ETP job training programs, which try to minimize red tape, do not reimburse trainers until all enrollees are placed in jobs. Some of the Silicon Valley community colleges are reluctant to become involved in these efforts because of this "performance-based" requirement. However, several community colleges adept at working out such contracts have begun ETP-sponsored programs.

COOPERATIVE INDUSTRY–COMMUNITY COLLEGE EFFORTS IN THE BOSTON AREA

In the Boston area, there are pockets of constructive relationships between the community colleges and high technology firms. As in

Silicon Valley, some of the schools have been involved in contract courses with a specific company whereby the college provides the upgrading training of some employees. Wentworth Institute, the oldest (founded in 1904) and largest (approximately 3,000 students) of the nonproprietary private two-year technical schools has also provided many such courses over the years. Seven of the community colleges, in cooperation with member firms of the Massachusetts High Technology Council, were also involved in CETA training programs in electronics technology. Middlesex Community College conducts a successful technical writing software program, which was originally aimed at laid-off schoolteachers. It was partially funded by federal occupational education funds with curriculum and instructional help coming from high technology companies, particularly Data General, Honeywell, and Digital Equipment.

The state-funded Bay State Skills Corporation has also brought together several community colleges and high technology companies in joint training efforts. The Corporation was the brainchild of George Kariotis, Governor Edward King's secretary of economic affairs and a former high technology executive himself. The state legislature set up the Corporation in 1981 and gave it $3 million over a two-year period, to be matched by industry funds, to develop programs in high growth fields. The hope was that educational institutions at all levels (including four-year colleges and universities) would plan courses of study and then approach companies for matching aid and, in some cases, for students. The effort has been successful and state funding for the program has continued. Dozens of programs have been funded, training almost 5,000 workers by 1984, many of them in high technology fields.

But some schools have been slow to develop proposals for funding from Bay State Skills, and many companies have been reluctant to come up with cash. And other schools sometimes have difficulty gearing up on short-term notice for one-shot programs. A number of firms shy away from any program funded through a government agency. But as of 1984, companies had come through with $3 million in contributions of various sorts. Response to the Corporation has been more positive among both industry and education people than was the case with the CWETA program in California. Senator Paul Tsongas used this program as a model in proposed legislation that would establish a national industry-education consortium (the U.S. Skills Corporation) that would draw in state, federal, and corporate funds to train people in high growth industries. Several states have already replicated the Massachusetts program.

A number of colleges have active industry advisory boards, some of which work well and some of which do not. A common complaint among community college officials was the lack of money to hire staff who could nurture those advisory relationships. Most of the two-year schools have at least one member from high technology firms on their boards of trustees. But few companies have donated equipment or have given other kinds of contributions to the community colleges. A major exception to this is the $3-million gift of new computer equipment by Wang Labs to each of the state's public institutions of higher education. All of the recipients acknowledged that the gift was a significant addition to their facilities. Digital Equipment has donated loaned personnel and equipment to Roxbury Community College and Middlesex Community College. And the Massachusetts High Technology Council has designed an electronics curriculum guide for technician training programs.

Overall, however, firms have donated fewer resources to the community colleges than their corporate counterparts have in Silicon Valley. And the Boston area schools have been less likely to provide on-site training for the companies. It may be that the Private Industry Councils that were set up or expanded under the federally funded Job Training Partnership Act in 1984 will facilitate the development of joint training programs between community colleges and firms.

In a number of ways, then, the organization and curricula of community colleges in both regions do reflect the needs of high technology companies, and a number of ties exist between the two sectors. The corporate managers interviewed in both places believed that the community colleges were making a real effort to respond to the needs of the work force in the high technology sector. Many claimed that of all sectors of education, the community colleges were most responsive to industrial shifts. The Boston area managers, though, usually added that there were several colleges that have done little in the way of technology-oriented training. (The one Massachusetts community college that does specialize in technical training, Springfield Technical Community College, is located outside the belt of high technology companies.) Most who were familiar with the two-year schools realized that the colleges were operating with less than adequate budgets and facilities. Silicon Valley corporations even contributed $70,000 to the unsuccessful 1984 campaign for a special tax levy to aid the Foothill-DeAnza Community College District.

College officials for their part expressed eagerness to offer programs that would prepare students for jobs in high growth companies. They felt that such jobs represented the wave of the future and believed a share of the school's resources should be committed to technical training. These administrators believed that they had the capacity to be flexible in the length and content of course programs. In Massachusetts, new programs can be developed through the self-supporting mechanism of continuing education, an enterprise enrolling approximately 50,000 students across the state.

Impediments to Industry–Community College Collaboration

Yet the relationship between the community colleges and high technology industry is an uneasy one, particularly in the Boston area where the community colleges do not have the size and status they have in California. The many impediments to collaborative programs are remarkably similar in both places.

An important obstacle to cooperation is one of educational philosophy: Many educators object to tying curricula too closely to the needs of one industry or one firm and feel that while the college should be attentive to economic trends, the schools should provide a range of programs that serve diverse needs. Community college educators are cautious about fashioning programs carefully tailored to a particular company's requirements because they feel that their personnel demands are volatile and unpredictable, and that a company's support for their programs is sporadic at best. Company officials generally favor a greater vocational emphasis for the community colleges, as opposed to liberal arts, and see the colleges' role as a provider of industry-specific training. These comments of the schools' administrators illustrate the tension that exists on the question of schools' ties to industry:

"Community colleges resist specific job skill training for a specific industry. We think of ourselves as an educational institution and not a training institution. An industry comes in and wants a program. Then the need dries up and we are stuck with the instructors and the machinery. We are now resisting that because we have been burned a few times. We want people to have a broad outlook and broad careers. We are protecting community concerns by fighting off certain occupational programs that might overproduce personnel in a field or place them in narrow jobs." Dean of occupational education at a community college adjacent to Silicon Valley

"Industry puts pressure on us to expand in electronics. We have to study whether there is permanency to what we get involved with. We can't lay off workers like industry can. Industry doesn't realize that when we gear up for electronics, we have to increase all support staff. It is such a changing field that an institution that wants to be stable like a college can't get too close." Vocational education administrator of a Silicon Valley community college district

"Education has to be especially careful at the college level. Companies will drop things and leave education and a program hanging. The market goes soft and they leave you. The companies are out for themselves and will use a school to their own advantage. I don't fault them—that is the way industry is and should be but education must be wary of them. I believe in education. It comes closer than any other institution to the 'pursuit of truth.' Industry is a bitter place to be." Administrator of a technical vocational program at a Silicon Valley college

"I am always fighting attempts to make this a technical institute. We want to leave some room for the liberal arts and have a significant percentage of our students transferring to four-year colleges. We want some of our graduates to get high tech jobs but their training should not be as narrow as it is in some states. Businessmen are not people to worship. The only way to be successful in industry is to be very self-interested." President of a Boston area community college

"There is a built-in domination by the private firm when companies contract with community colleges for a training course on the company's site. The instructor can't bring up the issue of health hazards and labor unions and sometimes cannot give the broad theoretical training that will make an employee flexible. There are inherent conflicts with contract courses although there is an advantage to the student to take the course on-site." Director of continuing education at a Boston area community college

Educators' attitudes toward industry have an ambivalent quality: On the one hand, they are critical of industry's motives and reliability but, on the other hand, they are still eager to accommodate corporate needs.

Company officials often expressed a different view about the role of community colleges. The vocationalism that they now eschew in high school programs is espoused for post-secondary education:

"We want 'rifle sho+' training from the community colleges without general education courses. Then training programs could be completed faster. The community college degree programs are slow and arduous—we need a fast track program with no general education. Employees don't want broad programs. They don't have the time. Many of them are working parents." Corporate training manager at a Silicon Valley electronics firm

"The community colleges are good here but their programs are disproportionately weighted to courses that are not applicable to industry. They are driven by the needs of students and professors and not by needs of industry. They need to shift resources into vocational education more." President of an industry trade association in Silicon Valley

Corporate managers are irked by the bureaucratic procedures and slow reaction time of the community colleges, as they are with other segments of the educational system. Commenting on the development of collaborative training courses, one training manager remarked: "They always say 'we can't do it' because they haven't done it before. They have a fear of getting things passed through committee." Another training director claimed that "Community colleges are responsive to some of our specific needs but their reaction time is slow. Our requirements emerge rapidly and disappear." They feel that professors have too much autonomy, a factor that can block the development of a new program.

This perception was verified by college administrators who acknowledged that "we can't react fast enough to industry partly because faculty are attuned to the semester system and are often unwilling to change to short courses." Another noted that "There is a problem with faculty autonomy—we can't force an instructor to teach a course." Another pointed out that "teachers won't set up new programs unless they want to. College administrators don't develop programs—they are developed by teachers." Schools also move slowly in hiring full-time faculty. These decisions may take as long as a year. In contrast, high technology companies change personnel and policies rapidly. In one company, the personnel director hired engineers in such quick fashion that he was forced to rely on telephone calls and telegrams rather than letters. It is no surprise that executives in these businesses become frustrated with what they perceive as the glacial pace of educational change.

While the companies are bothered by the slow reaction time of the schools, the colleges have difficulty with the frenetic pace of the firms. The organization of the companies themselves is often in a state of evolution. Firms merge or are bought out, new ones appear overnight, and product lines appear and disappear. "Things seem to happen by accident in the companies," observed one Silicon Valley community college administrator who works closely on training programs with a number of electronics firms. "*Some* planning must be going on but I don't see it."

The problem of high turnover of personnel in high technology firms was felt more strongly at the community college level than other sectors of education. "Training directors turn over like popcorn" said the director of one technology division in a Silicon Valley community college. "We are forever educating the personnel at one local company. There are always new people there who call and ask the same question: 'Do you have any programs in electronics that could be of use to us?' I burn up when I hear that." The dean of science and technology at a Silicon Valley community college pointed out that in the previous year there had been four different contact people at one firm who had replaced one another. In one employee upgrading program taught on-site at a company by a Boston area community college, the firm's coordinator changed three times in one year. "We might as well pencil in their names," complained the electronics instructor. Many in both industry and the community colleges reported that when they tried to develop a collaborative program, it fell through because the company contact changed jobs. The turnover rate "must drive the colleges bananas" acknowledged one training official at a semiconductor company. Complaints about changes in personnel were more numerous among educators in Silicon Valley, where turnover rates are higher, than among their Massachusetts counterparts.

Collaborative training relationships are sometimes impeded by companies' ambivalence about having their employees receive further training on the job or in the community colleges. They want to retain their workers, which often requires opening up new career paths for them, but then fear that upgrading will cause these same employees to jump to a better paying job at another company. Line supervisors resent losing workers for released-time courses. During periods of expansion, employees are often required to work overtime and have to drop out of night classes.

There are some firms that take a longer-range view and offer employees incentives (released-time) and rewards (promotions) for completing an entire degree program or a certificated course of study. However, many firms give very little in the way of encouragement in this regard. Also, the short-term interest of some companies in Silicon Valley during boom times leads them to hire students away from community college programs before they have completed the course of study. This trend helps account for the fact that the great majority of students who take courses in technical areas in the Santa Clara County community colleges do not receive a two-year degree. Fear of companies' hiring students prior to graduation led one college to forbid corporate recruiters from interviewing their electronics technology students. One company offered to hire all the graduates in the electronics technology program of one college sight unseen, an indicator of the intense demand for trained personnel in this field when the economy is in an upturn. The firms in the Boston area generally have a lesser demand for technicians and do not put such employment pressures on students.

Community college officials believe there is an elitist attitude on the part of most companies who focus their attention on programs with four-year colleges and universities. "Companies are more concerned about engineers than technicians, which shows an intellectual snobbery," complained the director of a Silicon Valley technology program at a community college. "We try to get high level company people on advisory committees but it is hard to get them. They see it as a lowly technology program." Community college officials also complain that few companies will allow them to come on-site and use the firm's equipment for training purposes, equipment that the colleges cannot afford to buy for themselves. A recent survey done of the National Association of Manufacturers bore out this contention that business people are reluctant to allow their equipment to be used by public vocational education training programs. Almost three-fourths of the nearly 800 companies responding to the survey said they would not be willing to have their equipment used for this purpose. [19] College administrators also complained that high technology companies were not good about taking work-experience students.

Both industry and education people talked about the day-to-day difficulties of developing partnerships between their respective institutions. The additional work load such cooperative efforts require deters many from taking on such projects, particularly since both institutions are short on personnel. "It took fifteen meetings to organize one contract course with a computer company, a huge amount of time. There were bureaucratic obstacles in both organizations, and nitpickers on both sides," commented an electronics instructor at one community college. "We made 45 persentations at one computer company in an effort to develop training programs but our proposals never got off the ground," lamented one Boston area college official. "I have enough problems with my own bureaucracy without having to deal with the bureaucracy of the community colleges. I'm so busy now—in the future I will coordinate with them," explained a training manager of a semiconductor firm.

A government relations manager in another company argued that "People in education forget there's a recession in the companies. Our manpower is stretched thin. When an employee works with them, that's in addition to his own work. They don't consider our needs." The president of a Boston area community college pointed out that "Our relations with community colleges could be better but we can't afford to put someone on full time to interact with companies. We haven't done a good job of letting companies know what we can do but we are very understaffed." In addition, many educators cited the difficulty of finding the right person to approach in a firm.

Overall, community college administrators in both the Silicon Valley and Boston areas felt that high technology industry gave them little support. Interestingly, even though California college officials do in fact receive considerably more in the way of donations and training contracts than their Boston counterparts, their attitudes were just as critical and pessimistic. Some of the following comments show their feelings on this question:

"Companies have generally not come through with equipment and often don't even return our phone calls. Only one company has been generous to us. And many of the high tech companies don't want the help in training. Once a company hires a training director, all is lost." President of a Boston area community college

"Equipment donations are nowhere near the amount we need. Some firms are quite generous but many others donate obsolete equipment and then complain that students are not trained on new equipment. There are almost no cash donations from industry." Director of vocational education at a Silicon Valley community college

"Budget cuts have hit us very hard and we are in a sliding mode downward now. More and more of my time goes into soliciting money and equipment. Some companies are donating useful equipment which came after a lot of pleading on my part." Chairman of the technology division of a Silicon Valley community college

"Industry, with a few exceptions, has no long-term commitment whatsoever. They clamor when there is a time of need but when no need exists, they give us the cold shoulder. They are concerned only with their immediate needs. Industry doesn't understand its own problems let alone education's problems." Director of engineering and technology at a Silicon Valley community college

"Industry's number one priority is making the buck and nothing else matters." President of a Silicon Valley community college

"Business gives very little in any way." Dean of applied sciences at a Silicon Valley community college

"We were encouraged by a high government official to develop a jointly funded training program in electronics. I wrote 60 high technology people and then called 14 personally to come to a reception. Only 6 or 7 showed up to hear our appeal for $25,000 in matching money for the training program. One company finally gave $5,000. I am somewhat bitter." President of a two-year technology college in the Boston area

Thus, the community colleges and local high technology companies in Santa Clara County and the Route 128 area around Boston have not developed partnerships with relative ease. Important and growing collaborative programs exist but the relationship is not as close as it is often reputed to be by public and private policymakers.

The Industry–University Connection

High techology firms and other corporations devote more time and money to fostering productive relations with university and four-year colleges than they do to building ties to community colleges and public schools. The shortage of newly trained engineering and computer science personnel and the desire to have access to emerging research developments in some technical fields have propelled companies to establish or deepen involvements with academic institutions. These involvements often include philanthropic gifts of cash or equipment or fellowship support, joint research efforts through individual research contracts or cooperative research programs, close recruiting relationships, membership in industrial affiliates programs, and personnel exchanges of various kinds.[20] Many companies have personnel assigned full-time to college relations and university recruitment. Most of the larger firms have established a nationwide program of "key schools" whereby they target their equipment donations, grants, and visits to schools that are most likely to meet their employment needs.

While neither educators nor industry people are fully satisfied with the relationship between the two sectors, the links are comparatively strong and still growing. The expansion of these connections occurs within a supportive political context since state and national governments see such collaborative efforts, particularly in research, as important to national economic productivity. The need for the United States to become more competitive in world markets is often cited as a reason why cooperative relations between research centers in firms and universities should be encouraged. Joint research projects, especially those that have the added benefit of training graduate students, are seen as a more efficient use of the nation's resources. It is interesting that the trend toward more cooperative company-academic work is occurring in other Western countries as well.[21]

University and college professors have long been hired as consultants by private industry, a phenomenon that represents a major link to the corporate world. Consulting relationships are especially strong in technical fields. An investigation in the late 1960s found that almost two-thirds of engineering professors were engaged in consulting activities, a considerably higher proportion doing so than their other academic colleagues.[22] Similar percentages were reported in 1975.[23] Consulting is not just limited to faculty in technical fields at the major research universities. A study by Frank and Edith Darknell shows that there is a good deal of consulting activity, for example, among engineering faculty at the California state colleges.[24] In the field of molecular biology, where there is a close connection between basic research and the possibility of development of commercial products, a large proportion of faculty consult for private biotechnology firms.[25] Consulting relationships are an important factor in the development of broader industry-academic research ties, a point emphasized in the 1982 National Science Foundation report on the subject:

A striking finding of the field study of university-industry research relationships [by Lois Peters and Herbert Fusfeld] . . . is that the initial impulse in the majority of the sampling of university-industry research relationships came from the university. At first glance this would seem a rather one-sided relationship. Yet closer scrutiny reveals that a significant proportion of the academic researchers pursuing these relationships had prior consulting or other employment relationships with companies. . . . We have a multi-stage series of relationships: company wants technical/scientific advice (general or specific) and seeks out professor; professor/consultant sees opportunities for research and initiates research relationship; company tracks and (maybe) utilizes the research and makes employment offers to the bright graduate students and postdoctorates working on the project. . . .[26]

This cycle, which often becomes repetitive, shows a process by which "resources, people, and information" are shared between companies and academe.[27]

INDUSTRY TIES WITH ELITE SCHOOLS

There are numerous examples of close ties between colleges and universities and high technology industry in Silicon Valley and the Boston areas. Industry ties are particularly strong at prestigious research universities. Indeed, the presence of Stanford University and MIT, discussed in Chapter Three, was and continues to be a crucial ingredient in the development of the two regional technical complexes. Frederick Terman, Stanford's former dean of engineering and the "father of Silicon Valley," became interested in developing science-based companies with close links to Stanford in the 1930s, to be what he referred to as a "community of technical scholars." Terman assisted in the formation of Hewlett-Packard in 1937 and he and the physics department also provided support in the founding of Varian in that same year. After World War II, when Terman became dean, he expanded the School of Engineering, encouraged faculty and industry people to pursue collaborative work, supported faculty in efforts to start their own companies, developed the Stanford Industrial Park, which rents to high technology firms, and began a part-time graduate program open to local company engineers. As provost and, later, as vice president of Stanford, Terman built up the chemistry department in a way that encouraged the formation of a new complex of companies in the area, specializing in biology and medicine. New biotechnology companies are still forming there, drawing on talent in the medical school and related disciplines.

Because of Stanford's intimate involvement in the creation and expansion of electronic and biomedical firms in the Valley, its relationship with industry is close and enduring. Three hundred and fifty employees from fifty area companies are enrolled in the Stanford Honors Co-op Program, a part-time master's program taught partly through an interactive television system located in the companies themselves. The firms pay their employees' tuition, double the normal fee, and provide released-time. Auditors and "non-registered options" (students who take the course including examinations, but receive no credit for it) are also allowed, for a fee, so the courses reach more than 2,000 students. Professors grumble about the additional work load these latter students generate and companies complain that so few of their employees are accepted into the highly selective Co-op program. The percentage of Stanford engineering graduate students who are supported by business has gone "up, up, up" in the words of one school official, so that now between 25 and 35 percent of the students have their tuition paid by industry.

In addition to ties with industry through faculty consulting, student hiring, and the Co-op program, Stanford runs an Industrial Affiliates Program, which brings in $1 to $2 million a year to the School of Engineering. In exchange for having a special recruitment relationship and access to information about ongoing research projects, companies donate an average of $10,000 a year to one of fifteen or sixteen subgroups of faculty. Each subgroup has some ten contributing companies associated with it, and the faculty can use the funds for whatever professional purpose they wish (equipment, travel, fellowships, etc.). The school also has an advisory board that includes top industry representatives whose function it is to review the school's overall program.

Stanford's new Center for Integrated Systems (CIS), a microelectronics center focused on the development of very large scale integrated systems, is being partially financed by corporate sponsors. Twenty companies from across the country (including Hewlett-Packard and Intel, both headquartered in Silicon Valley) each contributed $750,000 to help construct the CIS. Ongoing research in this area is now heavily supported by the federal government, but Stanford officials hope that eventually at least half of all research funds will come from industry.

Companies receive several concrete benefits from their sponsorship of CIS. They have special opportunities to observe the work of the Center's students and are given preferential treatment in recruiting them. Although research projects are approved by a committee composed only of Stanford personnel, CIS has a Sponsors Advisory Committee for technical guidance, which tries to tailor research to the needs of companies. Corporate staff on sabbatical leave from their firms can be temporary research associates of the lab. CIS will develop state-of-the-art educational materials that can be used in company training programs, and short courses and conferences based there are available to company personnel. Companies that are Affiliates but not Sponsors receive a "limited sampling of CIS activities."[28] The CIS director, John G. Linvill, believes that corporate contributions and the subsequent flow of highly trained graduates to the companies "will represent a new relationship between universities and industry, comparable in importance to that which developed between universities and the federal government following World War II."[29] A study completed for the National Science Foundation concluded similarly that CIS "symbolizes a new era of productivity in university research through unprecedented cooperation of academia, government and industry."[30]

MIT, in Cambridge, Massachusetts, also has a multiplicity of well-developed and mutually satisfying ties with high technology industry. Many of the Boston area high technology firms were founded by MIT graduates, and others (as many as 165 by 1968 alone) spun off from government-funded MIT programs and laboratories.[31] The school's faculty are involved as consultants or directors of many companies, and MIT provides the largest number of technical and professional employees with graduate degrees to member companies of the Massachusetts High Technology Council. The university has an active Industrial Liaison Program with over 270 company members (generating close to $6 million in revenue annually to MIT), which provides the firms with access to the institute's research and staff resources.

Other links between MIT and industry exist as well: approximately 11 percent of the research on campus is funded by corporations; companies have funded several endowed chairs; an industrial park, MIT Technology Square, adjacent to the campus has been in existence for two decades and further industrial development on university-owned land is being carried out; 16 percent of MIT students are in cooperative work-study programs; short courses, including video courses, are offered to industry personnel; a Management of Technology master's degree has been developed jointly between the Sloan School of Management and the School of Engineering; and a small, part-time master's program in the Department of Electrical Engineering and Computer Science has been started for people already employed in industry. While the size and structure of the latter program falls short of industry desires, it may be a first step in what might become a much larger effort to provide continuing education with credit for experienced technical professionals.

MIT is the recipient of corporate largesse more than any other engineering school in New England. As of 1982, for example, MIT received more than $20 million annually from corporations for research, more than triple the amount it received from the private sector in 1977. Some of these research connections include an $8.5-million microbiology project funded by W. R. Grace & Co.; a VLSI center (mostly government funded), which receives $750,000 over three years from nine electronics companies; and "Project Athena," a study of the computerization of curricula, which is funded by gifts of personnel, equipment, and money valued at $50 million from IBM and Digital Equipment. Thus, like Stanford and other elite engineering schools, the relationship between high technology industry and MIT is relatively harmonious and close. The two institutions respond well to one another's needs.

Harvard University also has important ties to technologically sophisticated firms, particularly in the area of biotechnology research. The university has entered into multimillion-dollar, long-term collaborative research relationships in the fields of genetics and molecular biology with chemical companies including Monsanto, Hoechst, and DuPont. The Hoechst agreement with the Harvard Medical School and Massachusetts General Hospital provides for an infusion of $50 million over ten years from the company, the largest amount in any industry-academic research project in the United States thus far. One biotechnology company, Biogen, is a spin-off from the university. Harvard has also developed a robotics laboratory with assistance from Data General, Automatix, and Bell Laboratories.

The various University of California campuses are involved in a jointly funded industry-academic microelectronics research and training program, Microelectronics Innovation and Computer Research Operation (MICRO). The impetus for the program, which includes a microelectronics center at the Berkeley campus, came from former Governor Jerry Brown and executives at Intel, National Semiconductor, and Advanced Micro Devices, all Silicon Valley firms. MICRO has an advisory board composed of equal numbers of state government, university, and electronics industry representatives, although research proposals are reviewed by an all-faculty committee. Collaborative research projects developed by firms and faculty receive state funding to match that of the corporations. The program also includes money for graduate fellowships in microelectronics and computer science. This publicly supported research and training effort parallels similar initiatives in North Carolina, Minnesota, and Massachusetts, all states that hope to expand their high technology manufacturing base. The University of California at Berkeley itself, adjacent to Silicon Valley, is a major source of engineering and technical talent. More and more high technology firms are locating near the university and its engineering program has become more involved in graduate training of working engineers on company sites.

Generally speaking, most of the collaborative activity between companies and universities takes place between a small number of elite research universities and a few large corporations. The National Science Foundation report on industry-university research connections claims that "only a few industries, and comparatively few companies among them, pursue much research related contact with universities."[32] The report notes that in both academia and the private business sector, 80 percent of all research is done by 20 percent of the organizations.[33] The comprehensive study by Peters and Fusfeld reviews this concentration of research activity, pointing out that ten universities alone account for one-quarter of all federal research and development funds for campus research.[34] "Where elitism is mutually preceived," their report argues, "universities generally have little difficulty attracting a wide range of industrial funding."[35] Other types of interactions between schools and firms, such as recruiting relationships and industrial affiliates programs, are also concentrated among a fairly small number of institutions. In the field of engineering, the designation of a school as "elite" often includes large public universities with extensive engineering programs.

HIGH TECHNOLOGY PARTNERSHIPS
WITH OTHER UNIVERSITIES

Colleges and universities with less than elite status have fewer connections with high technology companies and receive considerably less from them. However, the concern about the engineering education crisis has caused industry trade associations and some individual firms to be more attentive to the needs of engineering and computer science departments in a broad range of colleges and universities.

A consensus among industry and educational leaders on the nature of the engineering education problem and some of its solutions was reached in the early 1980s. They agreed that engineering schools have critical problems, based fundamentally on inadequate financial resources, and have developed a list of ways in which corporations can come to their aid. The recommendations encourage industry to make donations of cash, equipment, and fellowships to the schools, loan instructional personnel, provide summer jobs and consulting opportunities to faculty, create more co-op placements for students, supplement faculty salaries, establish endowed chairs, and develop collaborative research efforts. The proposals they formulated also asked that industry support state legislative action both to upgrade salaries for public university engineering faculty and to increase state budget allocations for engineering and computer science education at these schools.[36] There are some notable cases of innovative and well-developed consortial arrangements between less elite schools and local high technology industry. George Mason University in northern Virginia is a good example of an institution that has developed a range of meaningful partnership programs with surrounding firms.

The American Electronics Association, the Massachusetts High Technology Council, and the Semiconductor Industry Association are making major efforts to identify the needs of engineering schools and to encourage business support of various kinds to those institutions.[37] The AEA and MHTC are cooperating in a campaign to get their member companies to donate 2 percent of their research and development budgets to engineering and computer science education. The efforts of the MHTC thus far have been quite successful with member firms' contributions surpassing the initial goals by a wide margin. The AEA's Electronics Education Foundation has raised money from several dozen companies to fund more than a hundred fellowships for doctoral students in engineering. Massachusetts electronics companies alone had contributed nearly $1 million to this fund by 1984. In addition, the association has succeeded in getting a bill passed in California that would give tax incentives to companies that give new engineering equipment to community colleges, an incentive that before had only applied to gifts to four-year schools. At the federal level, the AEA is lobbying for broader tax breaks on equipment gifts to higher education.

The Exxon Education Foundation is giving $15 million to sixty–six colleges and universities for teaching fellowships for engineering doctoral candidates and to a hundred departments of engineering and allied fields for salary supplements to junior faculty. Bell Labs has initiated a similar but smaller program. The Semiconductor Industry Association has funneled money from twenty-eight corporations into cooperative research efforts at several major universities through the Semiconductor Research Cooperative.

A few of the larger Massachusetts high technology firms have already been making contributions of various kinds to the state's engineering and computer science schools. Changes in tax laws have spurred the pace of donations. The Economic Recovery Tax Act of 1981 by the U.S. Congress included several provisions that give companies increased tax breaks for research and equipment gifts. Several firms have developed a reputation for relative generosity in giving—Digital Equipment Corporation, Analog Devices, Data General, Polaroid, Computervision, and Raytheon, among locally based companies, were mentioned most frequently by university administrators interviewed for this study. Only a few have established foundations—GenRad, Honeywell, and Polaroid. (General Electric, sometimes classified as a high technology firm, has long given through its GE Foundation.) While no college or university in Massachusetts enjoys corporate support to the extent MIT does, some others have received significant donations from high technology companies. Northeastern University has been the beneficiary of several major company contributions. Data General played a major role in helping the University of Lowell develop a computer science major by contributing equipment, loaning instructors, and advising on curriculum. And a group of six firms has aided Lowell in establishing a robotics laboratory.

The University of Massachusetts at Amherst has received significant help from Raytheon, Digital, Computervision, and Analog Devices among others. The university, for example, is one of fourteen with Digital in a program (PACE) that will develop new methods of computer instruction for nonengineering disciplines, a program in which Digital is donating $16 million worth of equipment to the schools. (Both Digital and IBM launched a series of ambitious donation and competitive granting programs with higher educational institutions in the 1980s.) UMass has also become a world leader in polymer research and has received support from eighteen firms for research initiatives in that area, including the development of the Center for UMass-Industry Research on Polymers.

Digital Equipment and UMass have also teamed up in a creative program to train mathematics and science teachers. Students who are selected for this special master's degree received undergraduate degrees in mathematics or science. They spend two summers and one academic year in the Math/Science Technology Education Project, which provides them with a teaching internship in a public school system, an internship at Digital, and formal course work in education and interactive technologies. Graduates then teach for three years, with some continuing to work part-time at Digital, and are then free to continue their employment in education or in industry.

Worcester Polytechnic Institute (WPI) has succeeded in garnering significant donations from industry for its research budget. Boston University, which only began soliciting corporate donations in the 1980s, has received substantial contributions from Digital, Analog Devices, and a number of other high technology firms. Both BU and the University of Massachusetts are among the twenty universities nationwide receiving $40 million worth of computer-aided design/computer-aided manufacturing (CAD/CAM) computers from IBM, the largest corporate donation in the company's history.

All of these schools provide in-service training for companies' technical professionals on and off company sites. Northeastern's part-time master's program in engineering reaches several thousand people a year and dwarfs all others in size, enrolling more than half of all part-time graduate engineering students in the state. It recently began an interactive video system of live courses hooked up to company sites, modeled after the program at Stanford. This program reached fourteen companies in 1984 and is "growing like wildfire" in the words of one university administrator. Northeastern also has developed a variety of short courses aimed at preventing obsolescence in skills among veteran engineers. In a bold move, it began to develop continuing education programs for engineers in Silicon Valley in 1985, and it may expand to other regions as well.

The University of Massachusetts at Amherst has a videotaped instruction program that enrolls more than 800 engineers in 100 company sites, and it is working on a sophisticated statewide telecommunications network that would bring two-way courses into workplaces. The university has also begun a master's degree program in engineering for practicing manufacturing engineers at a Digital Equipment plant in Springfield, Massachusetts. WPI offers two master's degrees in business and management via videotape for personnel in technically oriented companies. Both Northeastern and UMass are participant institutions in the newly formed National Technological University, which televises master's degree courses by satellite in computer engineering to corporations around the country.

Northeastern is also connected to high technology companies in pre-service training through placement of co-op students in positions in some of the firms. Some of the Massachusetts schools are also linked to the firms through industry advisory boards or through joint training programs developed with the aid of the Bay State Skills Corporation.

Another emerging area of industry-education collaboration centers on the development of a research and training microelectronics center in Massachusetts, a project of the Massachusetts Technology Park Corporation. At Northeastern's initiative, a consortium of industry, education, and government leaders have planned a $40-million center in Westborough, Massachusetts, that will train people from engineering and technician training schools as well as private sector employees in the design and production of semiconductor devices. Students from all of the state's nine engineering schools (three public and six private), will share the center's facilities. It is estimated that 1,000 students and company professionals involved in retraining will be serviced by the center annually.

The fact that such a diversity of schools in addition to the companies could overcome "turf" battles and cooperate on the microelectronics center venture is impressive. The directors of the center are to be drawn in equal numbers from industry, education, and government. Public funds for the project will be matched by donations of various sorts (equipment, cash, teaching personnel) from high technology companies. The Technology Park Corporation is also proposing the construction of a $40-million materials research center that would be financed in a similar manner. A biotechnology industrial park, the creation of the Worcester Biotechnology Development Corporation, is also planned around the University of Massachusetts medical school in Worcester, with the state donating a parcel of land for the company sites. Clark University and Worcester Polytechnic Institute are involved in a Technology Productivity Innovation Center, which will provide small businesses with technical and marketing services, a project receiving funding from the U.S. Small Business Administration.

In Silicon Valley, San Jose State University and the University of Santa Clara also have a variety of direct ties with local electronics companies. The engineering schools at both universities have advisory committees composed of high level company executives. The advice of these executives can be particularly valuable when new programs are being developed. San Jose State has been soliciting corporate donations for its engineering program, an effort that succeeded in raising $1.4 million in 1982 (compared with $70,000 in 1979). A Center for Productivity and Manufacturing Engineering, a CAD/CAM training facility, was established in 1983 at San Jose State with substantial help from Lockheed Missiles and Space Company of Sunnyvale and IBM, which has a large research facility in San Jose. IBM donated $2 million worth of CAD/CAM computers to the Center, part of its $40-million donation program in that field.

The university's School of Engineering has become more involved in the continuing education of engineers and now offers a part-time master's degree at company sites as well as nondegree seminars and workshops. The university is the major supplier of engineers to local electronics firms, and it has a reputation in the business world for making "substantial changes in the curriculum in tune with what happens here in Silicon Valley."[38] Its engineering faculty are involved in consulting relationships with local high technology firms.[39] Moreover, it leads all other state universities in California in industrially supported research and development, both in dollar amounts and as a proportion of all R and D expenditures on campus.[40]

The University of Santa Clara has a well-known part-time master's degree "early bird" program, running five days a week from seven to nine in the morning. More than 1,300 working engineers take the courses taught by 150 part-time faculty who also hold full-time jobs in industry. An ambitious renovation and construction of engineering education facilities is being partially financed by local industry contributions. San Jose State and the University of Santa Clara are sometimes included in companies' "key school" programs of outreach where fruitful contacts are cultivated for recruitment purposes.

Industry and Academe: Views of One Another

Industry executives interviewed were generally positive in their opinions of the college and university engineering programs in the two regions. They also gave high marks to some computer science programs but felt that others needed improvement in quality. Many were sympathetic to the difficulties university administrators face in reallocating scarce resources and were aware that the public colleges and universities were underfunded. "The first order of business [for higher education] is more money," observed one Route 128 chief executive officer. While they expressed disappointment at Stanford's reluctance to expand its Honors Co-op Program for working engineers and MIT's refusal to develop a large-scale part-time graduate engineering degree, at the same time they had grudging respect for the schools' policies on this issue.

But company officials also had criticisms. Most believed, especially in Massachusetts, that the state's four-year colleges and universities had an insufficient number of technology course offerings. However, after stressing they wanted to see more of a high technology thrust in the curriculum, they often then added comments favorable to the liberal arts. "I don't want a society dominated by scientists and engineers," said one. Many brought up the traditional business complaint that schools, with the exception of Northeastern University with its extensive student co-op program, are inadequately attuned to labor market trends and care little about student career placements. Some admitted that companies did a poor job of communicating their changing needs for personnel. "Academic institutions have no sense of urgency" about responding to changes in work force demands, argued one human resources administrator. He claimed, and others agreed, that while "industry moves at 60 miles per hour, the educational sector moves at only 15 miles per hour." Another Boston area manager, who had formerly been an administrator in the state public higher education system, spoke disparagingly of the bureaucratic maze that slowed down the adoption of new programs: "By the time a program is approved in public higher education, the need has gone or Northeastern has done it."

Other executives saw institutions of higher learning as "bumbling in their efforts" at approaching industry for help. One executive, for example, cited the failure of schools to write high quality, focused requests for company support. Others cited what they felt was the narrow arrogance of some schools who wanted corporate money but no advice on the development of computer science and other curricula. One electronics vice president commented that "Colleges are trapped in the mentality that 'we know what we're doing and we just want you to give us money.' " A manager in another company claimed that "Colleges think they know all the answers. They should consult us more about what we need. . . . Colleges see the companies as the poison sumac growing up around the mighty oaks." Some managers argued that many schools have failed to form viable employer advisory boards for fledgling programs. The training director of one large firm commented that those schools which were not arrogant often veered to the other extreme of "pandering" to companies' short-term needs. Others pointed out that no university has developed a systematic sequence of retraining programs that would prevent experienced engineers from becoming obsolete.

College and university officials generally have ambivalent feelings about their relations with the high technology sector. One university president observed that executives from electronics and computer companies were easier to work with than other business people because they are focused in what they want from the university and are idea-oriented and creative. They are "not very different from university faculty types," he said. He added that while they often have a narrow perspective on what they want ("more engineers, period"), they have also become broader in their understanding of the mission of a comprehensive university. Many administrators observed that several years ago high technology executives approached university relations primarily from the standpoint of pressuring the schools to produce more engineers. But university officials pointed out that business attitudes have changed so that they now ask how they can help engineering education.

Many university administrators interviewed in 1982 were moderately hopeful that some money would be raised through the "two percent solution" campaign of the AEA and the Massachusetts High Technology Council, and they were pleased that several executives were taking the lead in trying to convince the high technology business community to make contributions. Massachusetts public college and university administrators, like their community college counterparts, were also approving of the presence of high technology representatives on the Board of Regents. The industry advisory board of one public engineering school in that state has been instrumental in successfully lobbying the state legislature for more money for laboratory facilities.

With the exception of MIT and Stanford, however, college and university officials in Silicon Valley and the Route 128 area are generally critical of the level of high technology corporate giving to their institutions. "Companies here are young and are not yet socialized to the tradition of charitable giving," observed one Boston area university president. Many believed that the initial $14-million contribution goal of the Massachusetts High Technology Council fund drive (2 percent of R and D budgets) was set too low, significantly below the federal tax deductible limit of 10 percent of pre-tax earnings. (American corporations give, on the average, around 1 percent of pre-tax earnings to educational and charitable institutions; absolute amounts increased in the first half of the 1980s despite declines in business profits during that period.)[41] Most felt that with the exception of a small number of firms, companies give far too little. "Very very few of these electronics firms seem to come up with anything for the arts, the symphony or anything," lamented a board member of several San Jose arts organizations. "These new people in electronics firms don't know how to contribute. They're so into themselves and their million-dollar homes and their three

Mercedes."[42] Even the more generous electronics firms often give away less than 1 percent of their pre-tax net income.

Yet some of the schools acknowledged that they had only recently begun approaching companies for donations while others, especially Massachusetts state colleges, asserted they lacked the personnel to initiate contacts with the firms. The ties that did exist came largely through their adjunct faculty who held industry positions. Several college presidents who are experienced fund-raisers were generally despairing of the low level of financial support from all business sectors.

One university offical who has worked closely with high technology industry characterized school-business ties as "a struggling relationship." He and others cited several difficulties in developing collaborative efforts: the diverse nature of the companies and their frequent inability to agree on common policies; the fact that firms' educational efforts are usually fragmented and not fully institutionalized; the relative slowness of most companies to establish systematic and enduring recruiting efforts with the schools' new graduates (instead of constantly "pirating" personnel from other firms); and the paucity of companies interested in a sustained work-study/ cooperative placement program for students ("the most primitive level of cooperation"). Others claimed that the rapid growth rates of the companies made it hard to nurture long-lasting ties. "There growth is so frantic they don't even return my phone calls," complained one administrator. Several stressed that such ties, in order to be effective, had to be with high-level executives who had more influence and longer tenure in firms than middle managers.

Educators also pointed to industry ambivalence in seeking federal support for university programs, although industry associations have begun lobbying efforts for engineering education. Several university presidents commented on this issue:

"Industry is sticking its head in the sand by not going to Washington and talking about the national crisis in engineering education. It is a national problem and we have no national policies. . . . The companies believe in 'small government' and are thus caught on the horns of a dilemma." President of a Massachusetts college

"There is such ambivalence in the business community about government support, especially in very self-reliant companies. They see government as restrictive and don't feel the need for it. Many high technology companies haven't been dependent on government contracts." President of a major Boston area university

"Business is schizophrenic. They believe in less government spending but, at the same time, they want selective government support for certain functions." President of a major Massachusetts university

In fact, industry officials agreed that they generally do not feel comfortable with lobbying the federal government for educational aid. One manager in Silicon Valley pointed out that "Our company is conservative. It is hard for us to lobby for federal money for higher education even though in one of our divisions, 85 percent of our orders [for technical equipment] are from universities." A trade association official in Silicon Valley, in discussing the need for industrially supported university research, stressed that "We want a loose fellowship of vision, guidance and matching funds—a simple coordinated system of university-industry relations—nothing bureaucratic like NSF." Several educators claimed that it would take a crisis of some sort, such as losing a significant share of the minicomputer or software market to Japan, before business acted aggressively at the federal level on educational issues. "Companies really need to be scared before they will get their act together and lobby for federal support," commented the coordinator of a large industrial affiliates program.

Just as corporate officials sometimes criticized the schools for the long lag in their response to labor market changes, university officials faulted industry for having such a short-term employment perspective. They felt that more companies had not developed "mature" human resource policies that encouraged the systematic retraining and development of their experienced engineers throughout their careers. They argued that "high tech hires in fits and starts" with little planning beyond the next economic quarter. The uncertainties in the labor market for newly trained engineers, a problem discussed in Chapter Six, is an ongoing bone of contention between companies and the schools. The regular cycles of shortage and surplus characterizing this market have made administrators reluctant to be full allies with industry on employment and enrollment policy.

Some educational administrators felt they should speed up their institution's response times to employment trends, but others defended the long-range planning cycles of colleges and universities.

Faculty critics in Massachusetts, citing the fact that a small proportion (less than 3 percent) of the state's workers are employed as technical professionals in high technology firms, questioned whether dramatic curricular changes should be implemented to accommodate the needs of that modest share of the employment market. Economist Peter Doeringer and his associates have summarized the colleges' viewpoint well:

For a variety of reasons, existing institutions in the higher education sector cannot and should not respond fully to the shifting needs of the economy. . . . Substantial programs of basic research cannot be undertaken as short-term commitments, teaching is built around long-duration degree-granting programs; laboratory facilities cannot be built overnight; and state legislatures are not likely to provide the budgetary flexibility needed to underwrite such responsiveness.[43]

Despite these varying perspectives, however, industry and educational officials have reached a rough consensus that curricula of institutions of higher learning need to be weighted more toward scientific and technical subjects than they have been previously. Student demand, if nothing else, provides the impetus for program shifts.

High Technology Influence on the Massachusetts Board of Regents: A Case Study

In the early 1980s, one issue that came up repeatedly in Massachusetts in the mass media and in meetings of faculty, students, and other public officials was whether or not the Board of Regents of Higher Education was "controlled" by high technology executives. The *Boston Globe* referred to the "high tech entrepreneurs who now dominate the state Board of Regents" in a 1982 editorial entitled "High Tech's Higher Ed." Another article cited observers' worry that the "curriculum is being skewed to turn out qualified employees for high tech jobs at the expense of other disciplines."[44] A piece in the *Boston Phoenix* was entitled "Just Married: The High Tech/Higher Education Honeymoon."[45] Of the fifteen members of the original Board of Regents, created in 1980 when public higher education was reorganized, three came from high technology firms: David Beaubien, senior vice president of EG&G; Ray Stata, chief executive officer of Analog Devices and a founder of the Massachusetts High

Technology Council; and An Wang, founder of Wang Laboratories. In 1984, Stata and Wang were replaced by nonindustry people by Governor Michael Dukakis when their terms expired, but Beaubien was made chairman of the board.

Debates in the early 1980s about the future direction of the curriculum in public higher education generally ran to one of two extremes: one view painted public higher education, especially its tenured faculty members, as insensitive and unwilling to change curricular directions to meet market forces; the other view held that there is a coordinated effort by high technology firms, through their members of the Board of Regents, to turn the colleges and universities into a "training ground for Dr. Wang." Indeed, as an example of the latter view, one community college president interviewed said, "I resent governors who make college presidents turn institutions over to industry to train cogs in the wheel of American industry."

But the great majority of public college administrators interviewed believed that while the high technology industry executives on the board had more influence than others, they did not "control" the board. All agreed that the influence of the high technology regents on Governor Edward King (who appointed all original board members) was crucial in getting the governor to recommend a 10-percent increase in the 1982–83 budget for public higher education. Since the proposed budget increase was meant to cover faculty salary increases as well as the expansion or development of technology and health-related programs, administrators did not see the executives' support as being totally self-interested. Almost all favored the presence of high technology people on the board because they believed they were influential and forceful advocates of public higher education. They also were nearly unanimous in pointing out that after their appointments as regents, the executives had educated themselves about both the real budgetary problems of the schools and the schools' capacities to run solid programs.

The extremes of this debate show that, at times, "business and education are critical of one another without good foundation for their position," in the words of one high technology regent. For example, contrary to what many business people expressed in interviews, college enrollments in education dropped after the mid-1970s. A look at other enrollment figures in community college programs and those of public colleges and universities show that there have been many significant changes in students' curriculum choices in the last few years. Likewise, the charge by some faculty and students that Massachusetts public colleges and universities are being transformed into technical training institutes is ill-founded. In fact, the system until recently has had few technical training programs available to students, forcing them to attend more expensive private schools or proprietary schools if they wanted such training. The public engineering programs, for example, which students take for granted in midwestern and western states, have been in short supply in New England. While San Jose State University offers a full range of engineering majors to hundreds of students in Silicon Valley at a very low cost to the students, citizens who live in Boston have no local four-year public engineering school. The Massachusetts state colleges have been particularly lacking in technology-oriented offerings. Such programs at community colleges are still relatively new and small.

Thus, during the 1980s the regents have tried to build some balance into a system that has long denied many of its students access to programs that would lead to good jobs. Moreover, chairman David Beaubien is considered an advocate of a broad liberal arts education and Ray Stata, a leading spokesperson for high technology industry in New England, has written and spoken of the need for technical education to increase the humanistic content of its required courses.[46] Stata was also singled out by many of the educators interviewed as being particularly well-informed and attentive to the problems of individual schools, although they were quick to add that his long-range social vision was atypical of business leaders.

By 1983, criticism of the presence of the three high technology regents had subsided. Since the creation of the board itself, budgets for public higher education in Massachusetts have risen, in part because of the influence of these regents on the administrative and legislative process in state government. Because money for expansion of new technical programs was provided in these budgets, liberal arts subjects could operate without the massive cuts that would otherwise have been made to fund the new high-demand disciplines, such as computer science. Fears that the colleges would be turned into technical institutes have thus far not materialized. If anything, high technology industry through these regents has helped to shore up and improve an underfunded higher education system just as business executives (through the California Roundtable) have stepped in to support elementary and secondary schools, the weak point in California's education system.

Partnerships in Perspective

There is a rich and growing array of cooperative connections between high technology companies and academic institutions. While these links are closest and more numerous at elite universities, they exist at lesser-known schools as well. There is some reason to believe the optimistic projections proclaimed in the media that a new era of industry-education cooperation has begun. Yet, as we have seen, there are many problems in these relationships, particularly at the community college level and at nonelite colleges and universities. Moreover, attempts to create high technology industrial parks in conjunction with universities have thus far largely failed according to one 1983 study.[47] Even at the prestigious research universities, there are conflicts with firms over values, patent rights, and restrictions on information dissemination, among others.[48] The question of the freedom of investigators to pursue issues of their own choosing and at a pace uncompromised by commercial imperatives has become an important issue. And the troubling question of the institutional independence of academe is being raised by many observers, including historian David Noble and writer Nancy Pfund:

Universities in the United States have never been the autonomous, disinterested citadel of objective scholarship and social criticism that some lovers of learning imagine. . . . Nevertheless, the universities have provided a living for moderate dissenters, a vantage point from which to observe critically what is going on outside (if not inside) and a platform from which to address with relative safety controversial social questions. . . . Perhaps the greatest danger posed by the renewed industrial connection is the very real threat to this relative independence at a time when we need to rethink fundamentally the central economic and political questions of modern industry and democracy.[49]

It is important, of course, to keep in mind that industrial support of all kinds for higher education is still fairly modest. Academic administrators, by and large, remain skeptical (but hopeful) about the degree to which company donations of resources will be forthcoming and the depth of the long-term corporate commitment to collaborative efforts. And business commitments are not the only ones that are uncertain. Unpredictable political forces and events could also reawaken antibusiness sentiments on compuses.

Further, all serious researchers and observers of campus-corporate ties stress that private sector funds cannot begin to make up for any significant reduction in federal or state funding. As the Peters and Fusfeld study notes, "industry funding [for research] itself is based upon the existence of a stable university research community, and this in turn depends today on a substantial level of support from the federal government." The government, they point out, provides for the "underlying technical infrastructure of university research capabilities" without which industrially supported research could not function.[50] Similarly, the adequate operation of engineering and other kinds of technical education must depend on public funding at the state and national level. Thus, if higher education is to continue to move in a direction consonant with economic trends, increased government support toward that end will be essential.

8 Education, the Economy, and the Future

When Sputnik was launched by the Soviet Union in 1957, the U.S. government initiated major programs to revive American education. In the 1980s, as microelectronic devices transform the economy, the federal government has done little to revitalize a flagging educational system. Clearly some sectors of the educational system are out of synchrony with the growing technological sophistication of American industry. Public schools do not yet have the capacity to respond adequately to demands for a more technically trained labor force, even in such regions as Silicon Valley and the Boston area where students have a reasonable probability of technical employment. And while higher education is generally reorienting itself in a way that brings it into closer correspondence with the "new information society," even there, numerous factors inhibit its ability to adapt to a different economic environment.

At the public school level, the shortage of qualified mathematics and science teachers (and newly trained competent teachers of all subjects), the outdated science curricula, and inadequate budgets for supplies and equipment are hindering the development of first-class educational programs. Attempts to upgrade these programs are focusing primarily on the secondary schools, ignoring the general lack of science education in the elementary schools. Even in the secondary schools, the national discussion of educational improve-ment has yet to affect mathematics and science education at the lo-cal level. A 1984 survey of high school counselors and science and mathematics teachers in forty-three states found only 23 percent who said that new science and mathematics intitiatives to encourage students in those fields were underway in their schools.[1]

The recent action in most states to elevate high school graduation requirements and public college and university entrance standards to include more academic courses will increase students' exposure to mathematics and science, but the underfunding of some of these ex-panded programs and the assignment of many ill-qualified teachers to them is blunting the effectiveness of reform efforts. The action of many states, including California, to increase educational funding has brightened the situation somewhat, but thus far the additional monies fall short of what is needed. In higher education, there has been considerable publicity about shortages of faculty in technical fields and the need for new equipment, but solutions are not yet in place. Large numbers of qualified students are still turned away from four-year computer science and engineering programs because of inadequate funds for teaching and research facilities, and because of the small number of qualified faculty available for teaching. The same holds true for many community college systems.

High technology firms have become much more active in pro-viding resources of various kinds to some educational institutions to help alleviate the bottlenecks that have developed in the technical labor supply. There has been a noteworthy increase in cooperative research relationships, donations of equipment and personnel, par-ticipation in advisory boards, and other sorts of resource exchanges. It is ironic, however, that the budget cuts that have driven resource-poor schools to seek out private support simultaneously undermine the condition of equality required in truly successful partnerships between public education and private industry. In addition, most of these connections are happening in higher education and, within that educational sector, at the most prestigious research universities. For the most part, a small number of companies are engaged in the bulk of partnership activities. Clearly business involvement alone cannot be a substitute for a substantial infusion of public money if all schools and colleges are to be able to respond to a changing economic environment.

Pressures for Closer Alignment
of Industry and Education

Despite all the bad news about schools, there are many positive changes occurring. The quality of education has unquestionably emerged as a significant political issue in the mid-1980s, particularly at the state level. As unskilled manufacturing jobs shrink, public policymakers have become more concerned about job creation, focusing particularly on high technology industry and other high growth economic sectors. These political concerns about economic development are providing a powerful impetus for educational change. A new coalition of business and educational leaders, politicians and students is attempting to forge a more demanding and technically oriented academic system.

Such an attempt at realignment is consistent with prevailing views of the dynamics of educational change. Schools are responsive to economic forces in part because they are supposed to produce workers who have the capacities to fit into existing jobs. As work requirements are raised, students "invest" in appropriate kinds of education that will assist them in finding a job. As a result, discontinuities between school preparation and employer needs are reduced.[2] Moreover, when powerful business interests perceive that educational institutions are failing to teach students the attitudes and skills consistent with workplace requirements, they sometimes intervene more directly to reorient the educational system back into rough congruence with these requirements.[3] Current trends confirm that indeed both mechanisms for tying schools more closely to employer needs are at work in the 1980s.

There is clearly an effort afoot to have the educational system become more closely allied with what is perceived as an increasingly high technology economy. The frenetic activity within most states to strengthen teacher training and evaluation, and efforts in many to improve teacher salaries and in-service training, illustrate reform activity. Tightened graduation requirements, increased homework, and the movement for a longer school day and school year are aimed at raising skill levels. The increased funding for public schools and public higher education allotted in some states and under consideration in others, in order to retain and attract business, indicate that policymakers are interested in having education in greater harmony with economic trends. Burgeoning enrollments in post-secondary engineering, computer science, and technician training courses and growing enrollments in high school mathematics and science courses among the college-bound show that students are indeed "investing" in and responding to shifts in the job market.

There is ample evidence that influential business groups, including some in the high technology sector, are actively engaged in the movement for educational change in the 1980s. Although school-company connections remain fragmented, there is no question that direct and indirect ties have been revitalized in the late 1970s and 1980s. The Reagan administration, which favors reduced federal funding for education, has promoted programs whereby corporations donate personnel and equipment to schools. President Reagan even proclaimed the 1983–84 school year as the National Year of Partnerships in Education and set up a partnership office in the White House that disseminates information on successful collaborative efforts.

Federal legislation has mandated employer representative majorities on state vocational education councils and local and state councils that control federal job training funds. High-level corporate leaders sit on blue-ribbon educational commissions promoting educational change in the 1980s. Governors and legislators in many states have advisory groups that include business people to propose ways in which educational programs can become more adaptive to the employment needs of the private sector. In 1984, Texas businessman H. Ross Perot successfully spearheaded a major package of school reforms and a tax increase for public schools through the legislature. Statewide public higher educational governance bodies, including the powerful Boards of Regents in California and Massachusetts, have had leaders of high technology industry appointed as members. Collaborative research programs between companies and universities have been stepped up, and more and more business advisory boards for academic programs have come into existence, particularly in new fields such as computer science. "Adopt-a-school" programs utilizing hundreds of business volunteers in the public schools have taken hold in a number of American cities. Community colleges have made visible, concrete efforts to reorient their programs to serve corporate training needs by moving in a vocational direction in their regular curriculum offerings and by developing customized training programs for specific companies.

Business interests, then, have perceived that a crisis in education exists and have begun to intervene to reshape schools and colleges to meet corporate needs. Long-standing employer grievances about inadequate preparation of new workers have intensified, executives in certain high technology fields are feeling the negative effects of shortages of personnel in certain crucial skill areas, and the highly publicized quality of school systems of foreign competitors (Japan, West Germany, the Soviet Union) have generated heightened concern about the ability of the United States to compete in worldwide markets. These pressures appear to have prompted direct business involvement in educational change efforts.

Alongside the overt forms of intervention, the business community indirectly shapes educational directions through labor market demands. Schools exist within a particular economic structure and labor process that prevents them from becoming totally autonomous institutions.[4] Many students and parents are sensitive to job opportunities, especially in higher education, and request programs that will train them for employment. Educational institutions for their part do not want to be accused of being divorced from economic realities and make some attempt to conform to work force trends. This pattern of accommodation to employment markets is an ongoing process that can be relied upon to maintain the connection between educational goals and industrial technology.[5]

Interestingly, corporate views are no longer very different from those of most other interest groups involved in education. Converging viewpoints and mutually shared goals among those active in the school reform movement do not permit business to put its own distinctive stamp on schools.[6] Corporate insistence on vocational education in high schools has now been dropped in favor of a broad-based academic education, with vocational programs now being emphasized at the community college and four-year college level instead. The push for secondary students to learn communication, science, and mathematics skills along with problem-solving abilities and an aptitude for "learning how to learn," has replaced earlier corporate concerns with narrow vocational skills training, which is now inappropriate in such a rapidly changing work environment.

A variety of special commissions that include representatives from business, labor, and educational groups have arrived at a rough consensus on what is needed in schools. Their motives for change may differ, but the programs they advocate (higher teachers' salaries, stiffened academic course requirements, and so on) are frequently the same. The complaints of "liberal" college professors about students' deficient skill levels mirror those of "conservative" employers who bemoan similar inadequacies in their employees. Parents too voice criticisms of schools and students that sound very much like those of other groups. Thus, the corporate voice does not stand out in a distinctive way, but is only one in a chorus of complaints about the state of American education.

Parallels with the Past

Business involvement in school policies has occurred on a larger scale at an earlier point in America's history. During the closing years of the nineteenth century and the first two decades of the twentieth century, periods of major economic and social upheaval, newly formed large corporations entered into the fray of educational politics.[7] Big business, especially the National Association of Manufacturers (NAM), became extensively involved in the movement to "vocationalize" the schools, a movement which established the belief that schools should be explicitly linked to the labor market. Their efforts to develop vocational education in the nation's secondary schools, which culminated in the passage of the Smith-Hughes Act in 1917, included the adoption of formal standardized testing, guidance counseling, and the tracking of students into differentiated programs ostensibly based on intellectual ability. As is the case today, companies argued that skill levels on the job were being elevated, with many unskilled jobs being automated out of existence. The "Report of the Committee on Industrial Education" of the NAM in 1905 has language similar to that found in speeches and reports by business groups today:

In 1902 a contracting firm in New York City employed 4,900 skilled mechanics direct from Europe, paying them 50 cents per day above the union rate, because it was impossible to secure such valuable workmen in our greatest industrial center. We should not depend on Europe for our skill: *we must educate our own boys.*[8]

There are several other fascinating parallels between the current educational change movement and the politics of the vocational education movement. NAM feared industrial competition from Germany and from what it regarded as her superior capacity to educate and train skilled workers, a comparison that is still made today but extended to include schools in Japan and the Soviet Union. As in the 1980s, the movement for school change at the turn of the century included a diverse coalition of groups, including labor, manufacturing interests, social workers, educators, and political leaders. The motives and concerns of each group supporting vocational education varied then just as they do today among those interest groups actively seeking reform. Both then and now, industrialists predicated their argument for curricular and other changes on the notion that higher skill levels are needed among workers. Yet in both historical periods, large sectors of the labor force were and are experiencing a long-term deskilling of work as a result of technological innovations. Labor groups and some intellectuals in both periods warned that business involvement in education was too intrusive and extensive, threatening academic autonomy and freedom of thought.

Yet corporate involvement in schooling was far more extensive during the first few decades of the twentieth century than is the case today. Indeed, school administrators adopted "scientific management" techniques and spoke frequently of running their institutions as though they were business enterprises. Business and professional people dominated boards of education, and business leaders were seen as models to emulate.[9] It wasn't until the Great Depression that educators became disillusioned with business. A 1932 editorial in *School Life* entitled "Bad Business" put it this way:

By borrowing the terms of the market place, we tried to borrow from the temporary glory of the market place. We tried to improve education's estate by clothing her in scraps of royal purple snipped from the hem of the new king of America. Business was the undisputed monarch of America during the last decade. Let us talk no more of education as a business.[10]

Educational officials today have respect for certain aspects of business. But because the goals and organizational styles of the two sectors remain disparate, educators are wary and cautious in their dealings with companies. The nation has only recently emerged from a period of relatively strong antibusiness sentiment, quite unlike the context within which the vocational education movement flourished.

Schools and Business: Different Directions?

Although there is an effort underway to bring schools and high technology industry into greater alignment, it is unlikely that as complete a congruence as occurred in the early part of this century will emerge again. Changing social and political conditions have created new forces that preclude the predominance of business interests in determining educational goals and content. These forces, active today, counter corporate influence and reduce their effect on educational directions.

To begin with, the power of multiple interest groups in educational politics grew considerably in the 1960s and 1970s.[11] These groups include teachers' organizations, which have become powerful at the state level, associations representing the needs of certain groups of students (the handicapped, bilingual pupils), and groups organized to ensure equal educational opportunities for minorities and women and low-income students. Parents have mounted spirited campaigns to stop the closings of public elementary and secondary schools. Students at the secondary and college levels have organized in a vocal way around various issues affecting their rights. And courts have intervened in a wide array of educational controversies.

The California public schools, for example, were in a state of "shock and overload" between 1970 and 1980, as Michael Kirst has described it. The schools struggled to adjust not only to the addition of new programs, but also to a series of major external forces including enrollment declines, collective bargaining, desegregation, minimum competency tests, the movement for educational vouchers, Proposition 13, and the imposition of four different financial systems.[12] Each change that has shaken schools and colleges in the last two decades has involved new vested interest groups whose involvement in education is likely to remain. The days when a small number of interests, business or otherwise, could reshape schools have disappeared.

The traditional elites that heretofore have controlled public education have broken down and have not been replaced by any one identifiable group. Such fragmented governance has led to a situation where there is a "crossfire of conflicting demands" where "no one is in charge."[13] In the nineteenth century, school leadership was dominated by agrarian lay groups of Protestant, Anglo-Saxon men whose commitment to the common school movement resembled a religious crusade. They were replaced by a reform-minded cadre of professional educational "experts" in the first half of the twentieth century, "social engineers who sought to bring about a smoothly meshing corporate society," and who adopted the language and models of big business.[14] Since the 1960s, however, there has been a proliferation of groups active in educational governance. The complicated governance structure of contemporary public education itself has been described as "an organizational theorist's nightmare":

Rube Goldberg himself could not make an organization chart of the official—not to mention the private and informal—lines of authority, regulation, and accounting that now exist in American public education. . . . It would take a political scientist's lifelong work to disentangle even the local story.[15]

Another set of forces, these internal to schools, may blunt efforts by business or any other single-interest group to control or heavily influence educational processes, particularly at the public school level. The makeup of contemporary schools—the fact that they contain "captive populations" of students who frequently resist schools' efforts to teach them specific skills and values—can undermine the effectiveness of curriculum goals. Schools are also difficult to change because their goals are numerous and often contradictory.[16] And just because schools attempt to instill certain kinds of knowledge does not mean that they succeed in doing so since student motivation remains problematic. Moreover, the teaching profession has its own strong views about appropriate curricula, which makes them resistant to change from outside pressure groups.

Additionally, pressure from the high technology business sector to reorient secondary schools and colleges in a more technical direction is being challenged by economists and educators who argue that the makeup of the future labor force is unclear. Educators' initial acceptance of industry projections for large numbers of technically trained workers has changed to skepticism as government economists project relatively small employment markets in high technology fields. Educators can, with some justification, resist major changes in school programs without the possibility of dependable, accurate long-term work force projections. This resistance may be reinforced by the volatility in high technology industry itself. As the public watches the shakeout in the computer industry and the rapid changes in all kinds of technologies, public policymakers may decide that more time is needed to carefully assess trends. The fascination with and rush to adopt computer literacy programs in schools may abate somewhat as educators wait to see what kinds of computer skills are really needed by the majority of students. With the arrival of more "user friendly" machines on the market, for example, more school administrators are asking whether it makes sense to teach a significant percentage of students specific computer programming languages. The fact that waves of change follow so hard and fast on one another makes it easier for all interest groups to agree that core academic requirements at the secondary school level be maintained or upgraded.

There are also deep-seated long-term social changes which will inhibit the improvement in student achievement that corporations are promoting. Much has been written about the effects of television, the rise of two-income families, and the doubling of the divorce rate since 1970. While the effects of these changes on learning are not easily and simply documented, teachers and educational administrators feel strongly that such forces create new burdens for schools and for children. Curricula may be made more demanding, but if students complete their homework in front of the television or have fewer adults available to help them with it, the result of higher academic demands may be student frustration rather than accomplishment.

Most crucially, the additional public money that will be needed to fund many of the reforms being considered in the 1980s, especially the need to raise teachers' salaries, may not be forthcoming because of competing demands on the public dollar. At the state level, the costs of health care and other social programs are rising, a condition that will only become exacerbated as the population ages. The aging "infrastructure"—sewers, dams, bridges, roads—will also make heavy demands on state and municipal budgets. At the federal level, military spending at present or higher levels will preclude significant financial increases for education. Moreover, as more and more workers are displaced by automation and foreign competition, there will be efforts to have more of the educational dollar targeted for retraining older workers than for spending on traditional student groups. And the decline in the percentage of parents in the population reduces education's natural constituency. As David Tyack and

Elisabeth Hansot have put it, "It is easy to imagine a future in which community of commitment to public education atrophies, competition for scarce resources increases, and public schools endure a slow death, especially in those communities where the poor and minorities predominate."[17]

The operation of all of these forces will make it increasingly difficult to bring schools into closer harmony with current economic trends. Educational institutions are indeed semiautonomous organizations that do not respond to economic change in any simple and obvious way.[18] Even with the current surge of direct business involvement in all levels of schooling, combined with all the less direct labor force processes guiding student behavior, other crucial social, political, and economic factors affecting schools may cause current attempts to upgrade school achievement to founder.

Although the corporate voice may occasionally be paramount on a particular issue (such as school-related tax measures) in a particular state or city, its role is unlikely ever again to be as pervasive and powerful as it was earlier in this century. This is especially true at the public school level where there are a large number of significant actors setting educational agendas. When it comes to higher education, firms have greater influence. The connections between post-secondary institutions and business are stronger, and ties between both academic and vocational programs and specific labor markets are more direct. Industry's focus at this level is more relevant to the students who at this stage in life are closer to entry into the labor market. Moreover, business representatives have more direct influence over higher education where they are often a predominant force on boards of trustees. Colleges and universities are also more sensitive to the opinions of business people since they rely on them more for donations than the public schools do.

The merits of relatively loose and shifting ties between schools and corporations should not be minimized. Industry-education links can become so strong that they present a threat to schools' and universities' ability to maintain a critical objectivity and autonomy. Agribusiness, large chemical firms, and defense-oriented companies have essentially captured aspects of the research and training enterprise at some universities. The open discussion of research findings among academics involved in technical corporate ventures (especially in the field of biotechnology) has been muted by commercial considerations. Vocational training programs in secondary schools and community colleges have sometimes become too focused and atheoretical when linked too closely to the needs of one firm. The efforts of some trade associations and individual corporations to introduce a narrow and self-serving version of economics education into the secondary school curriculum is another example of unproductive school-business programs. It is essential for the free expression of ideas, the maintenance of the liberal arts tradition, and the development of critical perspectives that educational institutions keep some distance from private sector establishments.

On the other hand, educational institutions need some guidance and various other forms of concrete support in fashioning curricula that will prepare students for the contemporary labor market. No one wants schools to educate students in a way that will handicap them as they seek work. In some instances, this support is absolutely critical even during periods when school budgets are amply funded. For example, active vocational business advisory councils are needed to give accurate local work force projections and to advise on the content of specific secondary and post-secondary vocational programs, especially those in rapidly changing fields. The provision of cooperative work placements for vocational students and the use of company equipment on-site when its cost is prohibitive for schools and colleges are also essential. Philanthropic donations of equipment to vocational schools, colleges, and universities in particular instances can make a major difference in the development of a computer science or other technical program. Without some of the equipment donations we see currently, a number of teaching and research programs would not exist. Generally speaking, it is best if schools can have adequate budgets that enable them to purchase or lease equipment themselves rather than having to take on the time-consuming and often degrading task of soliciting corporate donations. But in some specialized fields, donations or use of company facilities are the only realistic way schools can offer state-of-the-art training.

At this point in time, companies can play a critical role in elementary and secondary schooling by providing a new constituency in support of public education. This support must include lobbying for additional public funds when needed and a willingness to criticize educational practices when the educational community is unwilling to reform itself. There is a movement by business leaders in some cities and states across the country to become part of what

Atlanta school superintendent Alonzo Crim calls a "community of believers," an involvement that implies a belief that schools and their students can have a better and more meaningful educational experience.[19] Urban schools especially need to know that there is hope for change and that the business community will participate in efforts to upgrade programs.

Other kinds of corporate support can be useful for various sectors of education but are not essential. Training programs for teachers, loans of company personnel for teaching purposes and curriculum development, involvement in the creation of magnet schools, and stepped-up career awareness programs are all examples of ties that can be beneficial but are not critical to the existence of good educational programs. Even corporate research relationships with colleges and universities, while frequently useful to both parties, are not necessarily crucial to the existence of top-flight educational and research efforts in academe.

A delicate balance needs to be struck—one in which schools and firms interact to meet mutual needs in specific areas, but one also characterized by separate and distinctive goals, organizational configurations, and financial structures. Ideally, industry and education should pursue parallel but separate paths, intersecting only in specialized areas where employers provide expertise and resources that are unavailable from public sources.

Policy Issues

Difficult questions confront government and education leaders as they try to plan for life in a new industrial era. They find themselves needing to revamp or expand educational programs with decreasing resources to carry out that task. In many parts of the nation, educators will continue to be preoccupied with the effects of declining enrollment and see it as the primary problem on the agenda. The unpredictability of future labor market needs, which make long-range planning a risky enterprise, will provide an easy rationale for avoiding true reform. Our knowledge about the net effects of job loss due to the application of new technologies at work is still very incomplete, and the research documenting the degree to which the deskilling of work is occurring rests on an even sparser scholarly base. Regional variations in educational trends, demographic shifts, political environments, and employment markets play havoc with any attempt to formulate national policies and models.

Other troubling issues intrude into the debate over the appropriate response of the educational system to the growth of high technology industry. Human service workers and advocates in states fear that any growth in education budgets, propelled by economic development concerns, will come at the expense of programs for the poor and disabled. (At the state level, after all, there is no military budget to cut.) And within education, liberal arts educators resent major reallocations of resources favoring technical programs. Further, the rivalry among states on the question of plant location may in the end be self-defeating from a regional or national perspective.

Disillusionment with aspects of high technology industry itself have led some groups to avoid close cooperation with it. Some citizen, labor, and academic organizations have raised objections to the environmental impacts (mainly water pollution) of high technology firms, levels of worker safety, and the nonunion labor practices of the industry. The role of modern electronics in the development of more lethal conventional and nuclear weapons has become a significant and controversial political issue in some localities. Many educators and students are hostile to firms that are deeply engaged in research and development of such weapons. All of these questions sharpen policy deliberations about the direction education should move in the closing years of this century.

Yet it appears that a diverse coalition of educators, politicians, business people, and parents, each operating with somewhat different goals, are coming together to effect educational change in the 1980s. Individuals and organizations representing a wide spectrum of opinion now agree that the teaching profession is underpaid, undertrained, poorly appreciated, and inadequately evaluated. Many states are increasing salaries, creating career ladders, and stepping up evaluations and training opportunities. (Merit pay schemes are also being tried again, although the history of such compensation strategies shows that they almost always fail.[20]) Virtually all of the states have stiffened academic course requirements for high school graduation, reducing the number of electives students take. There are systematic efforts in a number of school districts to assign more homework to public secondary students (who spend an average of seven hours a week less on homework than their private and parochial school counterparts).[21]

The long-term implementation of these and other reforms will cost substantial sums of money. The great bulk of the activity aimed at revamping education is occurring at the state level where budgets, even in good times, are in precarious shape. Greatly increased federal support is essential to augment these state efforts. The National Science Board's blue-ribbon Commission on Pre-College Mathematics and Science and its 1982 Annual Report on University-Industry Research Relationships make clear the necessity of federal direction and financing of technology-oriented programs.[22] Physicist Gerald Holton, a member of the National Commission on Excellence in Education, states the case for federal leadership forcefully:

Our decentralized educational system has never and never will be able to make significant across-the-board changes within a five- to ten-year time frame in response to a national challenge if the leverage of the relatively small but vital federal contribution is not brought to bear, in terms of both planning and financing. Moreover, without strong national leadership, the impulse for reform can easily dissipate; in education, the time required to achieve and make visible striking improvements is much longer than the normal attention span that is common in America for a public issue.[23]

It is doubtful that schools and the corporate sector will achieve anything more than an uneasy alliance in the years ahead. The alliance around specific programs will be stronger at certain levels of higher education, such as in elite research relationships, and in educational programs in more pro-business conservative southern and southwestern states. But the divergence of goals and organizational styles existing between schools and business will continue to create obstacles to collaborative efforts. Companies will still threaten to leave states if they pass revenue-generating tax increases or pass laws giving workers additional protections and benefits, a condition that will continually fuel conflict between the public and private sectors.

Actual plant closings and work force reductions will inevitably generate antibusiness feeling among segments of the public, including teachers' organizations that see themselves as part of the union movement. The gap between the private affluence of successful entrepreneurs and the poverty of public programs will be a thorn in the side of educators who have chosen to work in public schools. The contrast between the get-rich-quick mentality of electronics firms, illustrated so graphically in the famed Silicon Valley life-style, with the more altruistic values found in education will inhibit the development of long-term partnerships. If the United States intervenes militarily in a war in a Third World region (e.g., Central America), and that war receives major corporate backing, specific industry-education partnerships may unravel as college students and faculty once again engage in protest activities as in the Vietnam era.

For the moment, however, corporations can play an important role in upgrading American education. It is clear that in some states increased educational funding will come only after organizations of business leaders press for it. While specific collaborative efforts in some areas may fall by the wayside or fail to develop altogether, the more important broad political support for education by powerful companies may still remain a significant force in educational politics. As we have seen in California and Massachusetts, intervention by business groups has provided essential help in shoring up the weaker financial rungs of their public elementary, secondary, and higher educational ladders.

Since microelectronic technology is only in its infancy, policy assessments reached now will doubtless yield to new realities in future decades. The impact of this new technology on employment and education will evolve over a long period of time, but certain issues will be of enduring significance: the socially destructive conflicts that arise when the gap between public and private resources is wide; the problems generated when there is a mismatch between employees' skills and employers' needs; the difficulties inherent in linking the two worlds of schools and firms in meaningful cooperative ventures; and the conflict that exists between the desire of educational institutions to maintain their integrity and independence versus their need to be financially viable and responsive to labor market forces.

Educators and public policymakers have trouble planning with confidence when the economic implications of the new and transforming technologies are still so uncertain. Yet there is now agreement among business, political, and educational groups that students are better off both as citizens and as workers if they have a broad-based, sound pre-college education that includes adequate training in mathematics and science. There is a consensus as well that those who choose to pursue technical careers must have access to high-quality programs in colleges and universities. But concurrence on how to pay for such programs still eludes policymakers. Thus far, funding from public sources falls far short as we enter an electronic future. And genuine obstacles between education and industry prevent significant infusions of corporate resources into schools and post-secondary institutions of learning. Despite the calls for reform that dominate the political airwaves, cutbacks and decline still haunt the hallways of American schools.

Appendix on
Research Methods

This investigation of the relationship between education and high technology industry centered on three main topics: the inclination and capacity of the schools to respond to a growing industry's demand for a better-educated labor force, the direct and indirect influence of high technology firms on educational policies; the nature of the ties existing between industry and education; and the factors that promote or hinder collaborative efforts between the two. It seemed clear from the start that the shape of industry-education ties would probably vary depending on the level of education being studied. The public schools (kindergarten through twelfth grade), community colleges and other two-year post-secondary schools, and four-year colleges and universities operate in varying social, financial, and political contexts. Thus, it is necessary to examine the relationship between each of these segments and the industry since it seemed probable that such a differentiated and complex educational system would interact with economic forces and institutions in varying ways.

This study draws on several types of information. The primary source of data is personal interviews I conducted with officials from education, industry, and government in Silicon Valley and the Boston area. I interviewed 105 people in Silicon Valley between January and July of 1981, and another 130 individuals in the Boston area between September 1981 and April 1982. These unstructured interviews lasted approximately one hour in length. Four-fifths of the respondents were interviewed in their offices and the rest were interviewed over the telephone. Approximately thirty follow-up interviews were conducted in both regions between 1983 and 1985.

In Silicon Valley, thirty-nine industry representatives were interviewed, mostly managers in personnel or training in seven of the ten largest companies in the Valley, all employing over 5,000 workers locally. Company officials from five medium-sized firms (employing between 500 and 3,000 workers locally) were also interviewed; the five firms were selected at random from a complete list of local high technology companies of that size. One small (under 200 employees) "start-up" company was studied as well. In addition, representatives from four industry trade associations were interviewed as were executives outside of training and personnel in the twelve targeted companies who were recommended as knowledgeable sources on the topic at hand. In the Boston area, thirty-six high technology industry managers, executives, and representatives of trade associations were interviewed. At least one official from each of the nine high technology companies that employed more than 5,000 people locally and at least one from eight different companies employing between 1,000 and 5,000 workers were included in the study. All firms contacted agreed to participate in the research.

Educators in a wide range of positions were interviewed. In Santa Clara County, fifty-nine people were successfuly contacted: administrators (usually the associate superintendent for curriculum) in ten public school systems including all of the high school districts in the heart of Silicon Valley; administrators in eight community colleges; three university deans of engineering; and seven vocational education administrators. Interviews also were conducted with two teachers' union officials, six local educational leaders prominent in the field of microcomputers, seven university professors knowledgeable about computer science and the use of computers in schools, five state education officials, two leaders of national professional associations, and three administrators of groups who coordinate industry-education activities. Four government officials and one community organizer in Silicon Valley were also interviewed.

In the Boston area, forty-two educators working with the public schools were interviewed, a more extensive survey than the one conducted among school officials in Silicon Valley. Because the Route 128–495 area electronics complex is more dispersed geographically than that in Silicon Valley, seven school systems with eight high schools were selected for study rather than trying to canvass all school districts in high technology areas. Six of the seven school systems were located in the cities and towns with the highest concentrations of high technology industry on Route 128 and the seventh system was located in a city where a high percentage of high technology professionals and managers resided. In each of the selected school systems, interviews were conducted with either the school superintendent or associate superintendent for curriculum, the chairs of both the mathematics and science departments at each of the high schools, and at least one vocational or industrial arts coordinator in each district. Several other miscellaneous school personnel (e.g., guidance counselors, computer specialists) were also contacted. Interviews were carried out with directors at three proprietary training schools and with administrators at six of the two-year colleges (including five presidents of those schools). The study included four university presidents, seven deans of engineering and computer science, and eleven other high-level university administrators and officials involved in industrial relations. Seven government personnel were contacted as well as six individuals who have been leaders of organizations that coordinate industry-education activities. Industry representatives were queried on personnel and training practices, links to educational institutions, work force needs, and assessments of the capabilities of local schools and colleges. Educators in both regions who were interviewed were asked about the status of programs that had some obvious link to the needs of a high technology economy. At the public school level, questions centered on the condition of mathematics and science teaching, technical vocational educational programs, and computer-related instruction. At the post-secondary levels, the research focused on the status of technician training efforts in electronics, computers, and other high technology fields; computer programming and computer science instruction; and engineering-degree programs. Questions about links with companies were asked of all respondents.

In addition to the interviews, several other sources of data were utilized in this study. I conducted a questionnaire survey of mathematics and science teachers in eight public secondary schools in the Route 128 area around Boston in 1982. All 241 of these teachers were contacted and 66 percent (158 people) filled out the questionnaire. Also, data on American high school seniors, drawn from a national sample of 28,000 students were obtained and reanalyzed from the *High School and Beyond* study of the National Center for Education Statistics. Finally, I assembled a broad range of documents, studies, and reports on various aspects of the research topic.

Notes

Chapter 1

1. U.S. Congress, Office of Technology Assessment, *Technology and Structural Unemployment: Retraining and Re-employment of Displaced Adult Workers*, Washington, D.C.: Office of Technology Assessment, 1985.
2. Samuel Bowles and Herbert Gintis, *Schooling in Capitalist America*, New York: Basic Books, 1976; Arthur G. Wirth, *Education in the Technological Society: The Vocational-Liberal Studies Controversy in the Early Twentieth Century*, University Press of America, 1971, 1980; Marvin Lazerson and W. Norton Grubb (eds.), *American Education and Vocationalism: A Documentary History 1870–1970*, New York: Teachers College Press, 1974; David Nasaw, *Schooled to Order: A Social History of Public Schooling in the United States*, New York: Oxford University Press, 1979.
3. Christopher J. Hurn, *The Limits and Possibilities of Schooling*, Boston: Allyn and Bacon, 1978.
4. David Tyack and Elisabeth Hansot, *Managers of Virtue: Public School Leadership in America, 1820–1980*, New York: Basic Books, 1982.
5. Barbara Lerner, "American Education: How Are We Doing?" *The Public Interest*, Fall 1982, pp. 59–82.

6. For a listing and review of these studies, see "Symposium on the Year of the Reports: Responses from the Educational Community," *Harvard Educational Review*, February 1984, pp. 1–31.

7. U.S. Department of Education, National Center for Education Statistics, Martin M. Frankel and Debra Gerald, *Projections of Education Statistics to 1990–91, Volume I*, Washington, D.C.: U.S. Government Printing Office, 1982, p. 3; Martin M. Frankel, "Projecting a School Enrollment Turnaround," *American Education*, August/September 1981, pp. 34, 35; U.S. Department of Education, National Center for Education Statistics, *The Condition of Education, 1983 Edition*, Washington, D.C.: U.S. Government Printing Office, 1983, pp. 5–7; U.S. Department of Education, *Prospects for Financing Elementary/Secondary Education in the States, Volume I*, Washington, D.C., 1982, pp. 28–31.

8. U.S. Department of Education, *Indicators of Education Status and Trends*, Washington, D.C.: U.S. Government Printing Office, 1985, pp. 20, 21.

9. U.S. Department of Education, The National Commission on Excellence in Education, *A Nation at Risk*, Washington, D.C.: U.S. Government Printing Office, 1983, p. 21.

10. U.S. Department of Education, *Prospects for Financing Elementary/Secondary Education in the States, Volume I*, Washington, D.C., 1982, p. i.

11. U.S. Congress, House Committee on Education and Labor, reprinted in *Education Times*, March 25, 1985, p. 2.

12. See Gilbert R. Austin and Herbert Garber (eds), *The Rise and Fall of National Test Scores*, New York: Academic Press, 1982, for a comprehensive review of the test score decline. See also the U.S. Department of Education, The National Commission on Excellence in Education, *A Nation at Risk*, Washington, D.C.: U.S. Government Printing Office, 1983, for a summary of changing patterns of achievement in American schools.

    The pattern of changing test scores varied by age group, subject area, region of the country, race, income group, and achievement level according to data from national surveys of nine-, thirteen-, and seventeen-year-olds conducted periodically by the National Assessment for Educational Progress since 1969. Elementary-aged students, through the fourth grade, have generally held steady or have improved in their achievement levels while upper-grade elementary pupils and junior and senior high school students have dropped in their performance on a variety of tests. See Barbara Lerner, "American Education: How Are We Doing?" *The Public Interest*, Fall 1982, pp. 59–82. Reading scores have not shown the pattern of decline that has characterized scores in all other subjects; indeed, nine-year-olds made significant gains in reading during the 1970s. Test levels in the South improved relative to other regions of the country in reading while the western states evidenced the sharpest relative decline in mathematics. Scores of minority, low-income, and low-achieving students have tended to rise in the 1970s and 1980s especially in reading among nine-year-olds and in math among thirteen-year-olds. But scores of high-achieving students of all ages have dropped, most notably in the areas of mathematics and science. See National Assessment of Educational Progress, *Three National Assessments of Science: Changes in Achievement, 1969–77*, 1978; *Changes in Mathematical Achievement, 1973–78*, 1979; *Three National Assessments of Reading Changes in Performance, 1970–80*, 1981; *Reading, Science and Mathematics Trends: A Closer Look*, 1982; Education Commission of the States, Denver.

The decline in basic skills in mathematics appears to have halted according to the 1982 NAEP survey, with nine- and seventeen-year-olds now holding steady and thirteen-year-olds rising in their scores. However, students do most poorly on sections involving higher-order problem-solving skills just as they do most poorly on portions of reading tests that examine inferential reasoning. Thus there appear to be improvements in more routine, easily learned subject matter, but poor performance in areas requiring more complex thinking persists. See National Assessment of Educational Progress, *The Third National Mathematics Assessment: Results, Trends and Issues*, Education Commission of the States, Denver, 1983.

13. U.S. Department of Education, The National Commission on Excellence in Education, *A Nation at Risk*, Washington, D.C.: U.S. Government Printing Office, 1983, pp. 1, 8,9, 19.

An Advisory Panel to the College Board and to the Educational Testing Service, in its analysis of trends in College Board Scholastic Aptitude Tests (SAT), has argued that the decline in SAT scores developed in two stages. It pointed out that most of the drop in SAT and similar test scores in the 1960s was accounted for by changes in the composition of the population taking the tests and planning to attend college. Many more lower-income students, minorities, women, and lower-achieving students became included in the test-taking group, a phenomenon that explains between two-thirds and three-fourths of the drop in scores between 1963 and the early 1970s. In the 1970s, however, a period of even more marked decline in scores, probably one-fourth of that decline could be attributed to compositional changes in the examined population. Scores of high achievers dropped during the 1970s, a trend not found in the earlier period. The Advisory

Panel concluded that deterioration in performance levels in the 1970s was due instead to "pervasive forces" in the school and the larger society. See College Entrance Examination Board, *On Further Examination: Report of the Advisory Panel on the Scholastic Aptitude Test Score Decline*, Princeton: CEEB, 1977. See also Bruce K. Eckland, "College Entrance Examination Trends," in *The Rise and Fall of National Test Scores*, Gilbert R. Austin and Herbert Garber (eds.), New York: Academic Press, 1982, pp. 9–34.

14. U.S. Department of Education, The National Commission on Excellence in Education, *A Nation at Risk*, p. 18; Clifford Adelman, National Institute of Education, "Devaluation, Diffusion and the College Connection: A Study of High School Transcripts, 1964–1981," commissioned by the National Commission on Excellence in Education, Washington, D.C., 1983.

15. Donald A. Rock, et al., "Factors Associated with Test Score Decline," Educational Testing Service, Princeton, 1984. See also U.S. Department of Education, National Center for Education Statistics, *High School and Beyond: A Capsule Description of High School Students*, Washington, D.C., 1981, p. 7; John C. Flanagan, "Analyzing Changes in School Levels of Achievement for Men and Women Using Project TALENT Ten and Fifteen-Year Retests," in *The Rise and Fall of National Test Scores*, Gilbert R. Austin and Herbert Garber (eds.), New York: Academic Press, 1982, p. 49; U.S. Bureau of the Census, "School Enrollment—Social and Economic Characteristics of Students: October, 1983," Washington, D.C.: U.S. Government Printing Office, 1984. There is some evidence from the 1983–84 National Assessment of Educational Progress that junior high and senior high school students are doing more homework than their counterparts in 1980: *Education Week*, May 15, 1985, p. 6.

16. College Entrance Examination Board, *On Further Examination: Report of the Advisory Panel on the Scholastic Aptitude Test Score Decline*, Princeton, CEEB, 1977; Bruce K. Eckland, "College Entrance Examination Trends," in *The Rise and Fall of National Test Scores*, Gilbert R. Austin and Herbert Garber (eds.), New York: Academic Press, 1982, pp. 9–34.

17. U.S. Department of Education, The National Commission on Excellence in Education, *A Nation at Risk*, Washington, D.C.: U.S. Government Printing Office, 1983, pp. 18–20.

18. Bruce K. Eckland, "College Entrance Examination Trends," in *The Rise and Fall of National Test Scores*, Gilbert R. Austin and Herbert Garber (eds.), New York: Academic Press, 1982, p. 24; W. T. Weaver, "In Search of Quality: The Need for Talent in Teaching," *Phi Delta Kappan*, September 1979, pp. 29–46; Gary Sykes, National

Institute of Education, "Teacher Preparation and the Teacher Work-force: Problems and Prospects for the Eighties," U.S. Department of Education, Washington, D.C., 1981.

19. California State Department of Education, *1979–80 Annual Report*, Sacramento, 1980, pp. 217–227; Donald A. Rock, et al., "Factors Associated with Test Score Decline," Educational Testing Service, Princeton, 1984.

20. U.S. Department of Education, National Center for Education Statistics, *High School and Beyond: A Capsule Description of High School Students*, Washington, D.C., 1981, pp. 17, 18.

21. See articles in special issue of the *Harvard Educational Review, Rethinking the Federal Role in Education*, November 1982. See also Donald A. Rock, et al., "Factors Associated with Test Score Decline," Educational Testing Service, Princeton, 1984.

22. For a review of changes in mathematics and science education, see Jane M. Armstrong, Education Commission of the States, "Results of a Fifty-State Survey of Initiatives in Science, Mathematics and Computer Education," in *Educating Americans for the 21st Century: Source Materials*, National Science Foundation, Washington, D.C., 1983, pp. 141–203.

23. George H. Gallup, "The Fifteenth Annual Gallup Poll of the Public's Attitudes Toward the Public Schools," *Phi Delta Kappan*, September 1983, pp. 33–47.

24. Center for Public Resources, *Basic Skills in the U.S. Work Force*, New York, 1982, p. iv.

25. National Science Foundation, *University-Industry Research Relationships: Myths, Realities and Potentials*, Fourteenth Annual Report of the National Science Board, Washington, D.C.: U.S. Government Printing Office, 1982.

26. Dale Mann, *"All That Glitters: Public School/Private Sector Interaction in Twenty-Three U.S. Cities,"* unpublished report for the Exxon Education Foundation, Teachers College, Columbia University, New York, 1984, p. 19.

27. *Education Week*, November 7, 1984, p. 13.

Chapter 2

1. Frederick L. Zieber, "Future Technology and Silicon Shock," DATA-QUEST, Conference on "Microelectronics in Transition: Industrial Transformation and Social Change," University of California at Santa Cruz, May 1983.

2. Gene Bylinsky, "Here Comes the Second Computer Revolution," *For-*

tune, November 1975; reprinted in *The Microelectronics Revolution,* Tom Forester (ed.), Cambridge: MIT Press, 1981, p. 5.

3. *San Francisco Chronicle,* September 22, 1980, p. 6.

4. Lynne E. Browne, "Commentaries," in *New England's Vital Resource: The Labor Force,* John C. Hoy and Melvin H. Bernstein (eds.), New England Board of Higher Education, Washington, D.C.: American Council on Education, 1982, pp. 93, 94.

5. *Business Week,* March 28, 1983, p. 85.

6. Massachusetts Department of Manpower Development, *Defining High Technology Industries in Massachusetts,* Boston, 1979; Massachusetts Division of Employment Security, *High Technology Employment in Massachusetts and Selected States,* Boston, 1981.

7. U.S. Bureau of Labor Statistics, Richard W. Riche, Daniel E. Hecker, and John W. Burgan, "High Technology Today and Tomorrow: A Small Slice of the Employment Pie," *Monthly Labor Review,* November 1983, pp. 50–58.

8. U.S. Census Bureau data cited in *Fact Book on High Technology and Energy-Related Higher Education in the West,* Western Interstate Commission for Higher Education, Boulder, Colo., 1983, p. 7; U.S. Congress, Joint Economic Committee, Robert Premus, *Location of High Technology Firms and Regional Economic Development,* Washington, D.C.: U.S. Government Printing Office, 1982, p. 6.

9. Candee Harris, "High Technology Employment and Employment Growth: Considerations of Firm Size, 1976–1980," unpublished manuscript prepared for the Industrial Science and Innovation Project, National Science Foundation, June 1984.

10. U.S. Department of Commerce, Bureau of Industrial Economics, *High Technology vs. Smokestack Industries,* Washington, D.C., 1983.

11. U.S. Bureau of Labor Statistics, Richard W. Riche, Daniel E. Hecker, and John W. Burgan, "High Technology Today and Tomorrow: A Small Slice of the Employment Pie," *Monthly Labor Review,* November 1983, pp. 50–58.

12. Marshall Goldman, "Industry Threats Can't Be Ignored," *Boston Globe,* May 29, 1984, p. 24.

13. *Business Week,* March 28, 1983, p. 87.

14. *Wall Street Journal,* February 14, 1983, p. 1.

15. *Boston Globe,* June 6, 1983, p. 1.

16. U.S. Congress, Joint Economic Committee, Robert Premus, *Location of High Technology Firms and Regional Economic Development,* Washington, D.C.: U.S. Government Printing Office, 1982, p. 23.

17. *San Jose Mercury News,* June 5, 1983, p. F1, in story written for the *Dallas Morning News* by Richard Alm and Karen Blumenthal.

18. *Electronic News*, August 16, 1982, p. 14.
19. *Wall Street Journal*, April 18, 1983, p. 1.
20. *San Jose Mercury News*, June 5, 1983, p. F1.
21. *Electronic News*, August 16, 1982, p. 15.
22. *Chronicle of Higher Education*, March 16, 1983, p. 19.
23. Henry M. Levin and Russell W. Rumberger, "The Educational Implications of High Technology," Institute for Research on Educational Finance and Governance, School of Education, Stanford University, 1983.
24. U.S. Bureau of Labor Statistics, Richard W. Riche, Daniel E. Hecker, and John W. Burgan, "High Technology Today and Tomorrow: A Small Slice of the Employment Pie," *Monthly Labor Review*, November 1983, pp. 50–58.
25. *Business Week*, March 28, 1983, pp. 85, 86.
26. *San Jose Mercury News*, March 13, 1983, p. 1.
27. Roy Rothwell and Walter Zegveld, *Technical Change and Employment*, London: Frances Pinter, 1979; C. M. Cooper and J. A. Clark, *Employment, Economics and Technology: The Impact of Technological Change in the Labour Market*, New York: St. Martin's Press, 1982.
28. *Business Week*, March 28, 1983, p. 86.
29. See four-part series by Pete Carey in *San Jose Mercury News*, January 8–11, 1984, p. 1. (Specific quote in January 9 issue.)
30. Tom Forester (ed.), *The Microelectronics Revolution*, Cambridge: MIT Press, 1981, p. ix, xiii–xvii.
31. Harley Shaiken, keynote address to Conference on "Microelectronics in Transition: Industrial Transformation and Social Change," University of California at Santa Cruz, May 1983.
32. Juan Rada, *The Impact of Microelectronics*, International Labor Office, Geneva, 1980, pp. 18, 19, 35, 36, 105; International Labor Office, *New Technologies: Their Impact on Employment and the Working Environment*, Geneva, 1982, pp. 6, 104; Paul Attewell, "Microelectronics and Employment: A Review of the Debate," Conference on "Microelectronics in Transition: Industrial Transformation and Social Change," University of California at Santa Cruz, May 1983; *Wall Street Journal*, June 23, 1983, p. 1.
33. U.S. Congress, Office of Technology Assessment, *Computerized Manufacturing Automation: Employment, Education and the Workplace*, Washington, D.C., 1984, pp. 101–176.
34. Harry Braverman, *Labor and Monopoly Capital*, New York: Monthly Review Press, 1974.
35. Robert T. Lund and John A. Hansen, *Connected Machines, Disconnected Jobs: Technology and Work in the Next Decade*, Cambridge;

MIT Center for Policy Alternatives, 1983; Diane Werneke, *Microelectronics and Office Jobs*, Geneva: The International Labor Office, 1983.

36. Henry M. Levin and Russell W. Rumberger, "The Educational Implications of High Technology," Institute for Research on Educational Finance and Governance, School of Education, Stanford University, 1983, p. 10.
37. Ibid., p. 5.
38. Ibid.
39. Paul Attewell, "The Deskilling Controversy," mimeo manuscript, Department of Sociology, State University of New York at Stoneybrook, 1982.
40. Paul Adler, "Does Automation Raise Skill Requirements? What, Again?" Conference on "Microelectronics in Transition: Industrial Transformation and Social Change," University of California at Santa Cruz, May 1983, p. 19.
41. International Labor Office, *New Technologies: Their Impact on Employment and the Working Environment*, Geneva, 1982, p. 163.
42. U.S. Congress, Office of Technology Assessment, *Computerized Manufacturing Automation: Employment, Education and the Workplace*, Washington, D.C., 1984, p. 110.
43. Center for Public Resources, *Basic Skills in the U.S. Work Force*, New York, 1982, pp. 13–15.
44. National Academy of Sciences, *High Schools and the Changing Workplace: The Employers' View*, Washington, D.C.: National Academy Press, 1984.
45. Michael Timpane, *Corporations and Public Education in the Cities*, unpublished report for the Carnegie Corporation of New York, Teachers College, Columbia University, New York, 1982; Leonard Lund and E. Patrick McGuire, "The Role of Business in Precollegiate Education," The Conference Board, New York, 1984.
46. Michael Timpane, *Corporations and Public Education in the Cities*, unpublished report for the Carnegie Corporation of New York, Teachers College, Columbia University, New York, 1982, p. 21.
47. U.S. Congress, Office of Technology Assessment, *Computerized Manufacturing Automation: Employment, Education and the Workplace*, Washington, D.C., 1984.
48. International Labor Office, *New Technologies: Their Impact on Employment and the Working Environment*, Geneva, 1982, p. 142.
49. U.S. Office of Education, National Commission on Excellence in Education, *A Nation at Risk*, Washington, D.C.: U.S. Government Printing Office, 1983, p. 13.

Chapter 3

1. Gene Bylinsky, "Boston: Cradle of Innovation," in *The Innovation Millionaires*, New York: Charles Scribner's Sons, 1976, pp. 73–91.
2. Gene Bylinsky, "California's Great Breeding Ground for Industry," in *The Innovation Millionaires*, New York: Charles Scribner's Sons, 1976, pp. 47–71; see also Everett M. Rogers and Judith K. Larsen, *Silicon Valley Fever*, New York: Basic Books, 1984, for a discussion of the development of the Silicon Valley and Route 128 complexes.
3. Christopher Rand, "Center of a New World," Part I, *The New Yorker*, April 11, 1964, p. 84.
4. Gene Bylinsky, *The Innovation Millionaires*, New York: Charles Scribner's Sons, 1976, p. 75; see also Christopher Rand, "Center of a New World," Parts I, II, and III, *The New Yorker*, April 11, 1964 (pp. 43–90), April 18, 1964 (pp. 57–107), and April 25, 1964 (pp. 55–129).
5. Gene Bylinsky, *The Innovation Millionaires*, New York: Charles Scribner's Sons, 1976, p. 53; Lenny Siegel and Herb Borock, *Background Report on Silicon Valley*, prepared for the U.S. Commission on Civil Rights by the Pacific Studies Center, Mountain View, Calif., 1982, pp. 13, 14; Massachusetts Division of Employment Security, *Business Support Services Employment in Massachusetts, 1975–1980*, Boston, 1982, p. 12; *Boston Globe*, February 16, 1984, p. 40; *New York Times*, February 20, 1984, p. A8; *San Jose Mercury News*, June 4, 1984, p. 6A.
6. *Boston Globe*, November 15, 1982, p. 1. From a three-part series by Ron Rosenberg comparing Silicon Valley with Route 128, November 14–16, 1982.
7. *New York Times*, February 6, 1984, p. 1.
8. K. Belser, "The Making of Suburban America," *Cry California* 5 (Fall 1970); see also Anna Lee Saxenian, *Silicon Chips and Spatial Structure: The Industrial Basis of Urbanization in Santa Clara County, California*, unpublished paper, Institute of Urban and Regional Development, University of California at Berkeley, 1981.
9. Gene Bylinsky, *The Innovation Millionaires*, New York: Charles Scribner's Sons, 1976, pp. 73–74.
10. Lynn E. Browne and John S. Hekman, "New England's Economy in the 1980s," *New England Economic Review*, Federal Reserve Bank of Boston, January/February, 1981, pp. 5–16.
11. Anna Lee Saxenian, "Silicon Valley: Regional Prototype or Historical Exception?" Department of Urban Studies and Planning, MIT, paper presented at Conference on "Microelectronics in Transition: In-

dustrial Transformation and Social Change," University of California at Santa Cruz, May 1983, Appendix A; U.S. Bureau of Labor Statistics, Richard W. Riche, Daniel E. Hecker, and John W. Burgan, "High Technology Today and Tomorrow: A Small Slice of the Employment Pie," *Monthly Labor Review*, November 1983, p. 56.

12. California Employment Development Department, "High Technology Firms and Employment in Santa Clara County, 1982," memo, San Jose, June 1983.

13. Massachusetts Division of Employment Security, *High Technology Employment in Massachusetts and Selected States: 1975–1981*, Boston, 1982; 1983 and 1984 figures on the percentage of high technology workers in the Massachusetts manufacturing work force were obtained from the Division of Employment Security. See also *Business Support Services Employment in Massachusetts, 1975–1980*, Boston, 1982. According to the U.S. Bureau of Labor Statistics' estimates, the percentage of all workers engaged in high technology industry in Massachusetts varied from 6.1 to 17.2 percent in 1982 depending on which of three definitions of high technology industry is used. Comparable percentages for California were 6.2 percent to less than 16.5 percent. See U.S. Bureau of Labor Statistics, Richard W. Riche, Daniel E. Hecker, and John W. Burgan, "High Technology Today and Tomorrow: A Small Slice of the Employment Pie," *Monthly Labor Review*, November 1983, p. 57; and James A. Parrott, et al., *Massachusetts High Tech: The Promise and the Reality*, Somerville, Mass.: The High Tech Research Group, 1984, for an extensive discussion of many aspects of high technology employment in Massachusetts.

14. *Boston Globe*, November 14, 1982, p. 1.

15. *San Jose Mercury News*, December 27, 1982, p. D1; John C. Hoy, "An Economic Context for Higher Education," in *Business and Academia: Partners in New England's Economic Renewal*, John C. Hoy and Melvin H. Bernstein (eds.), New England Board of Higher Education, Hanover, N.H.: University Press of New England, 1981, pp. 11–25.

16. *Boston Globe*, November 15, 1982, p. 1; *San Jose Mercury News*, December 27, 1982, p. D1.

17. *Boston Globe*, November 14–15, 1982, p. 1; For a comprehensive overview of Silicon Valley life-styles, see Pete Carey and Alan, Gathright, "By Work Obsessed," *San Jose Mercury News*, Feb. 17–23, 1985, p.1.

18. *Valley Living*, Sun Newspapers, April 1–2, 1980, p. A1; Everett M. Rogers and Judith K. Larsen, *Silicon Valley Fever*, New York: Basic Books, 1984, p. 29.

19. The San Jose SMSA had a divorce rate of 7.0 per 1,000 population in

1982 compared to 5.8 for California (1981) and 5.1 for the U.S. as a whole, according to the National Center for Health Statistics. It is tied for ninth on the list of SMSA's nationally for the highest divorce rate. The rate has declined somewhat since 1979. *San Jose Mercury News*, June 19, 1983, p. L1, See also Everett M. Rogers and Judith K. Larsen, *Silicon Valley Fever*, New York; Basic Books, 1984, for a description of Silicon Valley life-styles and values.

20. *San Jose Mercury News*, December 27, 1982, p. D1.
21. Ibid.
22. *Boston Globe*, December 23, 1982, p. 38. Citation of a study by Peat, Marwick Mitchell and Company's human resource consulting service in Boston.
23. Tracy Kidder, *The Soul of a New Machine*, Boston: Little, Brown, 1981.
24. U.S. Bureau of the Census, *General Population Characteristics: Massachusetts*, Washington, D.C.: U.S. Government Printing Office, 1982: *San Jose Mercury News*, January 3, 1983, p. 1, July 26, 1983, p. B1; June 11, 1984, p. 1; *Washington Post*, September 18, 1983, p. 1.
25. U.S. Bureau of the Census, *Advance Estimates of Social, Economic and Housing Characteristics: Massachusetts Counties and Selected Places*, Washington, D.C.: U.S. Government Printing Office, 1982; *Boston Globe*, October 16, 1982, p. 1; *San Jose Mercury News*, January 3, 1983, p. 1, February 19, 1983, p. A6.
26. *Boston Globe*, November 21, 1982, p. 23; *San Jose Mercury News*, April 13, 1984, p. 1A.
27. U.S. Department of Education, National Center for Education Statistics, *The Condition of Education, 1983 Edition*, Washington, D.C.: U.S. Government Printing Office, 1983, pp. 6, 7, 14, 26.
28. U.S. Department of Education, *Prospects for Financing Elementary/Secondary Education in the States, Volume I*, Washington, D.C., 1982, p. 29.
29. George Masnick, "Demographic Influences on the Labor Force," in *New England's Vital Resource: The Labor Force*, John C. Hoy and Melvin Bernstein (eds.), New England Board of Higher Education, Washington, D.C.: American Council on Education, 1982, pp. 36–64.
30. *San Jose Mercury News*, December 19, 1982, p. B1.
31. *San Jose Mercury News*, June 21, 1981, p. 9A, December 19, 1982, p. B1, April 8, 1983, p. B2.
32. U.S. Department of Education, National Center for Education Statistics, Lee R. Wolfe, *Revenues and Expenditures for Public Elementary and Secondary Education, 1978–79*, Washington, D.C., 1981.

33. Proposition 2½ limits property taxes to 2.5 percent of a municipality's full and fair cash value of property. It requires a municipality above this limit (some small towns were below already) to reduce its current levy by 15 percent each fiscal year until the levy is at the 2.5 limit. Once that limit has been reached, tax levies may increase a maximum of only 2.5 percent a year. During the first year of cuts (1981–82), many municipal officials avoided dramatic cuts in services by reevaluating property and receiving increased state aid. The Proposition 13 ballot initiative cut property taxes in half by putting a tax ceiling of 1 percent on the 1975 fair market value of property. Increases in assessed valuations are limited to 2 percent annually although when a home is sold, the new assessed valuation is the market price of the home.

34. Berman, Weiler Associates, *Improving Student Performance in California: Recommendations for the California Roundtable*, Berkeley, 1982, p. 59.

35. U.S. Department of Education, National Center for Education Statistics, *The Condition of Education, 1983 Edition*, Washington, D.C.: U.S. Government Printing Office, 1983, p. 22.

36. U.S. Department of Education, *Prospects for Financing Elementary/Secondary Education in the States, Volume I*, Washington, D.C., 1982, p. 52; National Education Association, *Rankings of the States, 1982, 1983*, NEA Research Memo, Washington, D.C., 1982, 1983; National Education Association, *Estimates of School Statistics, 1983–1984 and 1984–1985*, Washington, D.C., 1984, 1985.

37. California State Department of Education, *Student Achievement in California Schools, 1981–82, Annual Report*, Sacramento, 1982, pp. 153–186.

38. C. Emily Feistritzer, *The Condition of Teaching: A State-by-State Analysis*, Princeton: Carnegie Foundation for the Advancement of Teaching, 1983, p. 47; National Education Association, *Rankings of the States, 1983*, NEA Research Memo, Washington, D.C., 1983; National Education Association, *Estimates of School Statistics, 1984–85*, Washington, D.C., 1985.

39. Massachusetts Association of School Committees, "The Impact of Proposition 2½ on the Public Schools," Boston, 1982; Massachusetts Department of Education, "Report on the Effect of Proposition 2½ on Massachusetts School Districts, 1981–82," Quincy, Mass., 1983; C. Emily Feistritzer, *The Condition of Teaching: A State-by-State Analysis*, Princeton: Carnegie Foundation for the Advancement of Teaching, 1983, p. 37.

40. Massachusetts Department of Education, "Report on the Effect of Proposition 2½ on Massachusetts School Districts, 1981–82," Quincy, Mass., 1983.

41. *Boston Globe*, April 5, 1982, p. 13. By 1985, the number of resignations and retirements had nearly returned to their pre-Proposition 2½ levels.

42. U.S. Department of Education, *Prospects for Financing Elementary/Secondary Education in the States, Volume I*, Washington, D.C., 1982.

43. *San Jose Mercury News*, June 21, 1981, p. 9A.

44. *San Jose Mercury News*, June 24, 1981, p. 1.

45. *Washington Post*, September 18, 1983, p. A1.

46. *San Jose Mercury News*, January 12, 1983, p. B4, December 15, 1982, p. 1.

47. *San Jose Mercury News*, December 15, 1982, p. 1.

48. *San Jose Mercury News*, December 16, 1982, p. 1.

49. Massachusetts State Department of Education, "Report on the Effects of Proposition 2½ on Massachusetts School Districts, 1981–82," Quincy, Mass., 1983.

50. Lawrence E. Susskind and Jane Serio (eds.), *Proposition 2½: Its Impact on Massachusetts*, a report from the Impact: 2½ Project at MIT, Cambridge: Oelgeschlager, Gunn and Hain Publishers, 1983, pp. 141–148.

51. Massachusetts Association of School Committees, "The Impact of Proposition 2½ on Public Schools," Boston, 1982.

52. Rita B. Petrella, *Science Teaching in the Secondary Schools in Massachusetts, 1983*, Andover Public Schools, Andover, Mass., 1984.

53. California Teachers Association, press release, Burlingame, Calif., March 31, 1981.

54. *Education Times*, November 5, 1984, p. 7.

Chapter 4

1. U.S. Department of Education, "State Education Statistics: State Performance Outcomes, Resource Inputs and Population Characteristics, 1972 and 1982," Washington, D.C., January 1984.

2. Ibid.

3. U.S. Department of Education, National Center for Education Statistics, *The Condition of Education, 1983 Edition*, Washington, D.C.: U.S. Government Printing Office, 1983, p. 32.

4. Achievement scores from the 1980 *High School and Beyond* study survey of the U.S. Department of Education, National Center for Education Statistics, Washington, D.C., 1981, show New England students scoring several points higher than California students.

5. National Assessment of Educational Progress, *Three National*

Assessments of Science: Changes in Achievement, 1969–77, 1978; *Changes in Mathematical Achievement, 1973–78*, 1979; *Three National Assessments of Reading: Changes in Performance, 1970–80*, 1981; Education Commission of the States, Denver.

Nine-year-olds in the Northeast also lead other regions in reading scores although thirteen-and seventeen-year-olds generally rank second behind those in the Central states. Black students from the Northeast outscore black students from other regions at all ages in the reading assessments according to the NAEP studies cited above.

The Massachusetts State Department of Education has conducted several statewide tests with some of the NAEP items so that regional and national comparisons could be made. The results of those surveys conducted between 1974 and 1979 show a mixed pattern. The state's students scored above the national average in reading, mathematics, and some portions of the writing examination. In science, nine-year-olds also scored above the national average, but the rankings of the seventeen-year-olds placed them at the overall U.S. average. On some of the writing subtests, Massachusetts scored below regional and national averages. See Massachusetts State Department of Education, *Massachusetts Educational Assessment Program: Mathematics 1974–1975; Massachusetts Educational Assessment Program: Science and Ecology 1976–1977; Massachusetts Statewide Educational Assessment, 1977–78, Summary and Interpretations: Reading, 1978*, Boston. Analyses of test results from the state's mandatory testing of basic skills in 1981 showed that Massachusetts pupils are significantly weaker in writing than they are in reading and mathematics. See *Boston Globe*, October 28, 1981, p. 17.

6. California State Department of Education, California Assessment Program, *Student Achievement in California Schools, 1981–82 Annual Report*, Sacremento, 1982, pp. 139–143; *San Jose Mercury News*, November 17, 1983, p. B1, and May 11, 1984, p. 16A.

7. Karen Seashore Lewis, *The Quality of Public Education in Boston: An Assessment and Some Recommendations*, Center for Survey Research, University of Massachusetts, Boston, 1983.

8. Ibid., pp. 13–14.

9. *Boston Globe*, June 21, 1982, p. 1, and July 4, 1982, p. 1.

10. U.S. Department of Education, The National Commission on Excellence in Education, *A Nation at Risk*, Washington, D.C.: U.S. Government Printing Office, 1983, p. 19.

11. California State Department of Education, California Assessment Program, *Student Achievement in California Schools, 1981–82 Annual Report*, Sacramento, 1982, p. 170.

12. Ibid., p. 153.

13. California State Department of Education, California Assessment Program, *Student Achievement in California Schools, 1982–83 Annual Report*, Sacramento, 1983, pp. 173–175.

14. Ibid., p. 175.

15. National Science Board, Commission on Precollege Education in Mathematics, Science and Technology, *Today's Problems, Tomorrow's Crises*, National Science Foundation, Washington, D.C., 1982. See also National Research Council, *Indicators of Precollege Education in Science and Mathematics: A Preliminary Review*, Washington, D.C.: National Academy Press, 1985, for a comprehensive overview of the condition of science and mathematics education in the United States.

16. *Education Times*, July 9, 1984, p. 5, reporting a reanalysis of *High School and Beyond* data by the National Center for Education Statistics.

17. Kenneth Travers, Second International Mathematics Study, University of Illinois, reported in *Education Times*, October 8, 15, 22, 1984, p. 1; Marshall Smith, "Looking at U.S. Standing in International Math Achievement and the Education Reform Reports," *Education Times*, November 5, 1984, p. 2.

18. Willard J. Jacobson, Second International Science Study, Columbia University, 1984, reported in *Education Week*, June 6, 1984, p. 9.

19. Admissions Testing Program of the College Board, "National College-Bound Seniors, 1983," Waltham, Mass., 1983; National Assessment of Educational Progress, *The Third National Mathematics Assessment: Results, Trends and Issues*, Education Commission of the States, Denver, 1983.

20. *Education Week*, March 27, 1985, p. 6, reporting on a study by Market Data Retrieval, Westport, Conn.

21. Henry Jay Becker, "School Uses of Microcomputers: Reports from a National Survey," Issue No. 1, The Johns Hopkins University, Center for the Social Organization of Schools, April 1983.

22. *San Jose Mercury News*, May 7, 1982, p. B6.

23. California State Department of Education, California Assessment Program, *Student Achievement in California Schools, 1981–82 Annual Report*, Sacramento, 1982, pp. 189–200.

24. Santa Clara County Computer Education Consortium, "Santa Clara County Middle School Computer Curriculum Survey, 1982–83," Santa Clara County Office of Education, San Jose, 1983, p. 14.

25. Henry Jay Becker, "School Uses of Microcomputers: Reports from a National Survey," Issue No. 2, The Johns Hopkins University, Center for the Social Organization of Schools, June 1983.

26. U.S Department of Education, National Center for Education Statistics, *The Condition of Education, 1983 Edition*, Washington, D.C.: U.S. Government Printing Office, 1983, p. 172.

27. Ibid., pp. 177–178.

28. Gary Sykes, "Teacher Preparation and the Teaching Workforce: Problems and Prospects for the Eighties," National Institute of Education, U.S. Department of Education, Washington, D.C., 1981, p. 2.

29. Victor S. Vance and Philip C. Schlechty, *The Structure of the Teaching Occupation and the Characteristics of Teachers: A Sociological Interpretation*, unpublished paper, Campbell University and the University of North Carolina, 1982.

30. National Science Board, Commission on Precollege Education in Mathematics, Science and Technology, *Today's Problems, Tomorrow's Crises*, National Science Foundation, Washington, D.C., 1982; Sarah E. Klein, president of the National Science Teachers Association, "Testimony to Committee on Labor and Human Resources of the U.S. Senate," April 15, 1982; *Education Week*, May 23, 1984, p. 9, which refers to a 1984 study by Bill G. Aldridge and Karen Johnston, National Science Teachers Association, Washington, D.C. For an extensive review of this subject, see John L. Taylor, *Teacher Shortage in Science and Mathematics: Myths, Realities, and Research*, Proceedings of a conference sponsored by the National Institute of Education, Washington, D.C., 1984.

31. C. Emily Feistritzer, *The Condition of Teaching: A State-by-State Analysis*, Princeton, The Carnegie Foundation for the Advancement of Teaching, 1983, pp. 44–49, 112; National Education Association, *Rankings of the States, 1983*, NEA Research Memo, Washington, D.C., 1983; National Education Association, "Nationwide Teacher Opinion Poll, 1983," NEA Research Division, Washington, D.C., 1983; National Education Association, *Estimates of School Statistics, 1983–84*, Washington, D.C.

32. U.S. Department of Education, National Center for Education Statistics, *The Condition of Education, 1983 Edition*, Washington, D.C.: U.S. Government Printing Office, 1983, p. 176.

33. U.S. Department of Education, *Projections of Education Statistics to 1990–91, Volume 1*, Washington, D.C., 1982, p. 74; National Education Association, "Nationwide Teacher Opinion Poll, 1983," NEA Research Division, Washington, D.C., 1983.

34. National Education Association, "Nationwide Teacher Opinion Poll, 1983, NEA Research Division, Washington, D.C., 1983; *MTA Today*, Boston, January 26, 1983, p. 10.

35. James Shymansky, National Science Teachers Association,

Washington, D.C., 1982, cited by Sarah E. Klein, "Testimony to Committee on Labor and Human Resources of the U.S. Senate," April 15, 1982.

36. California Science Teachers Association, "Report on Science-Math Survey," unpublished, Berkeley, 1982.

37. James W. Guthrie and Ami Zusman, *Mathematics and Science Teacher Shortages: What Can California Do?* Institute of Governmental Studies, University of California at Berkeley, 1982.

38. Massachusetts State Department of Education, "The Supply and Demand of Mathematics and Science Teachers in Massachusetts," Quincy, Mass., 1983; C. Emily Feistritzer, "Teacher Education Reports," Vol. 6, No. 1, January 20, 1984, Washington, D.C.

39. Massachusetts Association of Science Teachers, Massachusetts State Department of Education, "State of Science Questionnaire," Quincy, Mass., 1982.

40. William Faulkner, Raymond Bisplinghoff, and Louis Klotz, New Hampshire Legislative Academy of Science and Technology, "Perspectives on the Teaching of the Sciences," Concord, N.H., 1982; National Education Association, *Estimates of School Statistics, 1984–1985*, Washington, D.C., 1985.

41. Massachusetts Division of Employment Security, *Occupations in Massachusetts: Projected Changes, 1980 to 1990*, Boston, 1983, p. 18.

42. California Employment Development Department, *Projections of Employment by Industry and Occupation, 1980–1985, San Jose SMSA*, San Jose, 1982, p. 32.

43. James W. Guthrie and Ami Zusman, *Mathematics and Science Teacher Shortages: What Can California Do?* Institute of Governmental Studies, University of California at Berkeley, 1982.

44. *San Jose Mercury News*, June 4, 1983, p. 1.

45. National Science Foundation, *What Are the Needs in Precollege Science, Mathematics and Social Science Education? Views from the Field*, Washington, D.C., 1979, pp. 47, 127.

46. Rita B. Petrella, *Science Teaching in the Secondary Schools of Massachusetts, 1983*, Andover Public Schools, Andover, Mass., 1984.

47. Christine LeCam, *The Public Vocational System as a Potential Source of Labor Supply to Selected High Technology Occupations*, Massachusetts Department of Manpower Development, Boston, 1980.

48. Patricia M. Flynn, *Production Life Cycles and Their Implications for Education and Training: A Study of High Technology in Lowell*, Bentley College, Waltham, Mass., 1984.

49. F. James Rutherford, "Commentary: Fun with Funding in Science and Math Education," *Education Week*, October 31, 1984, p. 18.

50. *San Jose Mercury News*, June 3, 1983, p. 7F, reporting on a Mervin Field poll of California residents conducted in March 1983.
51. *Washington Post*, September 18, 1983, p. 1.
52. *San Jose Mercury News*, July 19, 1983, p. 1, reporting on a Mervin Field poll of a representative sample of 1,516 Californians.
53. *San Jose Mercury News*, June 23, 1983, p. 1, reporting on a poll of 4,000 California residents commissioned by the Association of California School Administrators and the County Superintendents of Schools Association.
54. California State Department of Education, California Assessment Program, *Student Achievement in California Schools, 1982–83 Annual Report*, Sacramento, 1983, p. 186.
55. George Blakeslee, "Science and Mathematics Teacher Supply and Demand in the Competitive Professional Labor Market: A Systems Dynamic Policy Study," unpublished doctoral dissertation, Boston University School of Education, 1984.
56. Chris Pipho, "Loan Forgiveness Programs: Will They Work?" *Education Week*, April 17, 1985, p. 19.

Chapter 5

1. See, among others, Michael Katz, *The Irony of Early School Reform*, Cambridge: Harvard University Press, 1968; Samuel Bowles and Herbert Gintis, *Schooling in Capitalist America*, New York: Basic Books, 1976; David Nasaw, *Schooled to Order: A Social History of Public Schooling in the United States*, New York: Oxford University Press, 1979; and Arthur G. Wirth, *Education in the Technological Society*, Washington, D.C.: University Press of America, 1980.
2. Michael Timpane, *Corporations and Public Education in the Cities*, unpublished report for the Carnegie Corporation of New York, Teachers College, Columbia University, New York, 1982, pp. 24–26, 44–46.
3. New England Board of Higher Education, *A Threat to Excellence*, Report of the Commission on Higher Education and the Economy of New England, Wenham, Mass., 1982; James W. Guthrie and Ami Zusman, *Mathematics and Science Teacher Shortages: What Can California Do?*, Institute of Governmental Studies, University of California at Berkeley, 1982; California Commission on Industrial Innovation, *Winning Technologies: A New Industrial Strategy for California and the Nation*, Office of the Governor, Sacramento, 1982.
4. Task Force on Education for Economic Growth, *Action for Excellence*, Education Commission of the States, Denver, 1983.

5. Berman, Weiler Associates, *Improving Student Performance in California: Recommendations for the California Roundtable*, Report for the California Roundtable, Berkeley, 1982.

6. Joseph Bellenger, "The Interrelationships Between Education and the Electronic Industry of Silicon Valley," unpublished report, Office of the Santa Clara County Superintendent of Schools, San Jose, 1982, p. 3.

7. *Boston Globe*, November 14, 1982, p. 1.

8. Roland S. Barth, "Educators Possess the Power to Restore Themselves," *Harvard Graduate School of Education Association Bulletin*, Cambridge, Spring/Summer 1983, p. 18.

9. *San Jose Mercury News*, July 4, 1983, p. 2D.

10. Ibid.

11. *Boston Globe*, December 7, 1982, p. 32.

12. Michael Hilliard and James Parrot, *The Massachusetts High Technology Council: An Assessment of its Public Policy Agenda*, unpublished paper, Department of Economics, University of Massachusetts, Amherst, 1982. Contribution figures are provided by the Office of Campaign and Political Finance, Boston.

13. Eric Lundquist, "High Tech on the Hot Seat," *Boston Magazine*, February 1981, pp. 78–118.

14. *San Jose Mercury News*, October 15, 1983, p. 16D.

15. *New York Times*, June 26, 1983, p. F3.

16. *Business Journal*, June 6, 1983, reporting on a survey by the American Electronics Association.

17. Speech by Roderick MacDougall, Conference on "Striving for Excellence," Boston, November 21, 1983.

18. Joseph Bellenger, "The Interrelationships Between Education and the Electronic Industry of Silicon Valley," unpublished report, Office of the Santa Clara County Superindent of Schools, San Jose, 1982, p. 3.

19. Patricia M. Flynn, *Production Life Cycles and Their Implications for Education and Training: A Study of High Technology in Lowell*, Bentley College, Waltham, Mass., 1984.

20. *Boston Globe*, February 20, 1983, p. A34.

21. Michael Timpane, *Corporations and Public Education in the Cities*, unpublished report for the Carnegie Corporation of New York, Teachers College, Columbia University, New York, 1982, p. 24.

22. *San Jose Mercury News*, September 12, 1982, p. 1.

23. Ibid.

24. Berman, Weiler Associates, *Improving Student Performance in California: Recommendations for the California Roundtable*, Report for the California Roundtable, Berkeley, 1982.

25. Speech by Roderick MacDougall, Conference on "Striving for Excellence," Boston, November 21, 1983.

26. Joseph Bellenger, "The Interrelationships Between Education and the Electronic Industry of Silicon Valley," unpublished report, Office of the Santa Clara County Superintendent of Schools, San Jose, 1982, p. 13.
27. *San Jose Mercury News*, May 13, 1984, p. 1.
28. See Dale Mann, "It's Up to You to Steer Those School/Business Partnerships," *American School Board Journal*, October 1984, for a discussion of the weaknesses of urban school business partnerships. See also Dale Mann, "All That Glitters: Public School/Private Sector Interaction in Twenty-Three U.S. Cities," unpublished report for the Exxon Education Foundation, Teachers College, Columbia University, 1984.
29. Ian E. McNett, *Let's Not Reinvent the Wheel: Profiles of School/ Business Collaboration*, Institute for Educational Leadership, Washington, D.C., 1982; Marsha Levine, "Barriers to Private Sector/Public School Collaboration: A Conceptual Framework," in *Barriers to Private Sector/Public School Collaboration*, a set of exploratory papers commissioned by the National Institute of Education and the American Enterprise Institute for Public Policy Research, Washington, D.C., June 1983 (other papers in this collection also deal with common elements of success in partnerships); Commonwealth of Massachusetts, Executive Office of Economic Affairs, "The Report of the Governor's Task Force on Private Sector Initiatives," Boston, 1983; Elizabeth Useem, "Significant Issues Relating to Educational Technology in School-Business Partnerships," a review of papers on partnerships written for the Conference on "Education and Technology: Innovative Responses to Changing Needs," National Institute of Education, American Enterprise Institute, Washington, D.C., June 1984; American Council of Life Insurance, *Company-School Collaboration: A Manual for Developing Successful Projects*, Washington, D.C., 1984; U.S. Department of Education, *Partnerships in Education: Exemplary Efforts Across the Nation*, and *Partnerships in Education: Education Trends of the Future*, Washington, D.C., 1984.

Chapter 6

1. John C. Hoy, "An Economic Context for Higher Education," in *Business and Academia: Partners in New England's Economic Renewal*, John C. Hoy and Melvin H. Bernstein (eds.), New England Board of Higher Education, Hanover, N.H.: University Press of New England, 1981, pp. 12, 24; see also Patricia Pannell Flynn, "Finance

and the Future of Higher Education in New England," in *Financing Higher Education: The Public Investment,* John C. Hoy and Melvin H. Bernstein (eds.), New England Board of Higher Education, Boston: Auburn House Publishing Company, 1982, pp. 45–71.

2. D. Kent Halstead, *How States Compare in Financial Support of Higher Education, 1981–82,* National Institute of Education, Washington, D.C., 1982.

3. Ibid.; California Postsecondary Education Commission, *Postsecondary Education in California: 1982 Information Digest,* Sacramento, 1982, pp. 5, 6.

4. D. Kent Halstead, *How States Compare in Financial Support of Higher Education, 1981–82,* National Institute of Education, Washington, D.C., 1982; *Chronicle of Higher Education,* October 31, 1984, p. 1, reporting on a study by M. M. Chambers, *Appropriations of State Tax Funds for Operating Expenses of Higher Education, 1984–85,* National Association of State Universities and Land-Grant Colleges, Washington, D.C., 1984.

5. National Education Association, *Rankings of the States, 1983,* Washington, D.C., 1983.

6. *Chronicle of Higher Education,* October 20, 1982, and October 26, 1983, reporting on studies by M. M. Chambers of Illinois State University.

7. W. Norton Grubb, "The Bandwagon Once More: Vocational Preparation for High Technology Occupations," *Harvard Educational Review,* November 1984, pp. 429–451.

8. C. L. Alton, *An Advanced Technology Study for Post-Secondary Area Vocational-Technical Schools,* Georgia Institute of Technology, prepared for the Georgia Department of Education, Atlanta, 1982.

9. *San Jose Mercury News,* June 13, 1983, p. 3B.

10. American Electronics Association, *Technical Employment Projections 1983–1987,* Palo Alto, 1983, p. 10.

11. Coopers and Lybrand, *Results of 1982 Survey of Human Resource Needs,* Massachusetts High Technology Council, Boston, 1982, pp. 2, 3.

12. California State Employment Development Department, *Projections of Employment by Industry and Occupation, 1980–85 Update,* Employment Data and Research Division, San Jose, 1982; Massachusetts Division of Employment Security, *Occupations in Massachusetts: Projected Changes 1980–1990,* Boston, 1983.

13. The College Board, "National College-Bound Seniors, 1983;" "California College-Bound Seniors, 1983;" "Massachusetts College-Bound Seniors, 1983," Admission Testing Program, New York, 1983.

14. Engineering Manpower Commission, "Engineering Enrollment

Highlights, Fall 1982," *Engineering Manpower Bulletin*, New York, August 1983.

15. Engineering Manpower Commission, "Engineering Degree Statistics and Trends—1984," *Engineering Manpower Bulletin*, New York, December 1984.

16. Massachusetts Board of Regents of Higher Education, *Engineering Education: Current Trends and Future Directions*, Boston, 1984.

17. Joan H. Grebe, "Appendix: Engineering and Technological Education in New England," in *New England's Vital Resource: The Labor Force*, John C. Hoy and Melvin H. Bernstein (eds.), New England Board of Higher Education, Washington, D.C.: American Council on Education, 1982, pp. 131-149.

18. Coopers and Lybrand, *Results of 1982 Survey of Human Resource Needs*, Massachusetts High Technology Council, Boston, 1982, pp. 3, 13; California Postsecondary Education Commission, *Engineering and Computer Science Education in California Public Higher Education*, Sacramento, 1982, pp. 21-46; Western Interstate Commission for Higher Education, *Profiles: High Technology Education and Manpower in the West*, Boulder, Colo., 1983, p. 40.

19. *Chronicle of Higher Education*, November 25, 1981, p. 10; *Wall Street Journal*, January 14, 1983, p. 1; J. W. Geils, "The Faculty Shortage: The 1982 Survey," *Engineering Education*, October 1983, p. 47; U.S. Congress, Office of Technology Assessment, *International Competitiveness in Electronics*, Washington, D.C., 1983, p. 307.

20. *Chronicle of Higher Education*, November 25, 1981, p. 10, citation of a study conducted for the National Science Foundation by the American Council on Education. See also U.S. Congress, Office of Technology Assessment, *International Competitiveness in Electronics*, pp. 306-314, for an overall discussion of engineering education.

21. *Chronicle of Higher Education*, November 23, 1983, p. 19, reporting of a special survey conducted for the *Chronicle*. See also James Botkin, Dan Dimancescu, and Ray Stata, *Global Stakes: The Future of High Technology in America*, Cambridge: Ballinger Publishing Company, 1982, for a comprehensive discussion of the engineering education crisis.

22. Engineering Manpower Commission, "Engineering Degree Statistics and Trends—1983," *Engineering Manpower Bulletin*, New York, March 1984; "Engineering Degree Statistics and Trends—1984," *Engineering Manpower Bulletin*, New York, December 1984; and "Engineering Enrollment Highlights, Fall 1982," *Engineering Manpower Bulletin*, New York, August 1983.

23. *Wall Street Journal*, January 14, 1983, p. 1.

24. Pat Hill Hubbard, *Planting the Engineering Seed Corn*, American Electronics Association, Palo Alto, 1981; Engineering Manpower Commission, "Engineering Degree Statistics and Trends—1984," *Engineering Manpower Bulletin*, New York, December 1984.
25. *Chronicle of Higher Education*, November 25, 1981, p. 10.
26. Ibid.
27. Ibid., December 2, 1981, p. 3.
28. Ibid., November 23, 1983, p. 20.
29. Ibid., December 16, 1981, p. 7, and February 29, 1984, p. 17, citing a survey of the College University and Personnel Association.
30. Pat Hill Hubbard, *Planting the Engineering Seed Corn*, American Electronics Association, Palo Alto, 1981.
31. Massachusetts Board of Regents of Higher Education, *Engineering Education: Current Trends and Future Directions*, Boston, 1984.
32. *Boston Globe*, February 27, 1984, p. 1.
33. Ibid.
34. California Postsecondary Education Commission, *Engineering and Computer Science Education in California Public Higher Education*, Sacramento, 1982, pp. 47–56, 97, 98.
35. *Chronicle of Higher Education*, March 2, 1983, p. 1, reporting of testimony before the Science and Technology Committee of the U.S. House of Representatives.
36. U.S. Bureau of Labor Statistics, Douglas D. Braddock, "The Job Market for Engineers: Recent Conditions and Future Prospects," *Occupational Outlook Quarterly*, Summer 1983.
37. For a review of the debate, see U.S. Congress, Office of Technology Assessment, *Computerized Manufacturing Automation: Employment, Education, and the Workplace*, Washington, D.C., 1984, pp. 119–124. OTA predicts that demand for engineers will grow in the 1980s but that growth will slow in the 1990s in part because of automation. A Bureau of Labor Statistics report, however, predicts strong demand in some engineering fields through 1995. See U.S. Bureau of Labor Statistics, Tom Nardone, "The Job Outlook in Brief," in *Occupational Outlook Quarterly*, Spring 1984, pp. 3–25. See also reports of the National Science Foundation, which argue that the supply of engineers in the 1980s is now adequate to meet demand. National Science Foundation, *Science Resource Studies Highlights*, February 23, 1983, and Febrary 17, 1984.
38. Coopers and Lybrand, *Results of 1982 Survey of Human Resource Needs*, Massachusetts High Technology Council, Boston, 1982.
39. U.S. Congress, Office of Technology Assessment, *International Competitiveness in Electronics*, Washington, D.C., 1983, p. 309.

40. Engineering Manpower Commission, "Engineers' Salaries, 1984," *Engineering Manpower Bulletin*, New York, November 1984.
41. Joan H. Grebe, "Appendix: Engineering and Technological Education in New England," in *New England's Vital Resource: The Labor Force*, John C. Hoy and Melvin H. Bernstein (eds.), New England Board of Higher Education, Washington, D.C.: American Council on Education, 1982, p. 148. See also Massachusetts Division of Employment Security, Eugene Doody and Helen Munzer, *High Technology Employment in Massachusetts and Selected States*, Boston, 1981.
42. Engineering Manpower Commission, "Demand for Engineers 1982/1983," *Engineering Manpower Bulletin*, New York, June 1983.
43. American Electronics Association, *Technical Employment Projections 1983–1987*, Palo Alto, 1983, pp. 24, 39, 41.
44. Neville Lee, Bank of Boston, "The Demand for Engineers in Massachusetts," Massachusetts Board of Regents of Higher Education, Boston, February 1984.
45. American Electronics Association, *Technical Employment Projections 1983–1987*, Palo Alto, 1983, p. 7.
46. Paula Leventman, *Professionals Out of Work*, New York: The Free Press, 1981, an account of the impact of engineering layoffs on Route 128 in the early 1970s.
47. Engineering Manpower Commission, "Demand for Engineers 1983/1984," *Engineering Manpower Bulletin*, New York, May 1984.
48. American Electronics Association, *Technical Employment Projections 1983–1987*, Palo Alto, 1983, pp. 24, 90, 109.
49. Christine LeCam, *The Public Vocational Education System as a Potential Source of Labor Supply to Selected High Technology Occupations*, Massachusetts Department of Manpower Development, Boston, 1980.
50. U.S. Bureau of Labor Statistics, Tom Nardone, "The Job Outlook in Brief," *Occupational Outlook Quarterly*, Spring 1984, pp. 2–25.
51. U.S. Congress, Office of Technology Assessment, *Computerized Manufacturing Automation: Employment, Education, and the Workplace*, Washington, D.C., 1984, pp. 122–127.

Chapter 7

1. Lois S. Peters and Herbert I. Fusfeld, "Current U.S. University-Industry Research Connections," in *University-Industry Research Relationships: Selected Studies*, National Science Foundation, Washington, D.C., 1982, p. 21; National Science Foundation,

University-Industry Research Relationships, Fourteenth Annual Report of the National Science Board, Washington, D.C., 1982, p. 28.

2. *Business Week,* December 20, 1982, p. 58. See also Jana B. Matthews and Rolf Norgaard, *Managing the Partnership Between Higher Education and Industry,* National Center for Higher Education Management Systems, Boulder, Colo., 1984, and Ernest A. Lynton, *The Missing Connection Between Business and the Universities,* American Council on Education and Macmillan Publishing Co., New York, 1984.

3. Arnold Thackray, "University-Industry Connections and Chemical Research: A Historical Perspective," in *University-Industry Research Relationships: Selected Studies,* National Science Foundation, Washington, D.C., 1982, pp. 193–233.

4. David Noble, *America by Design,* New York: Oxford University Press, 1977.

5. National Science Foundation, *University-Industry Research Relationships,* Fourteenth Annual Report of the National Science Board, Washington, D.C., 1982, p. 4.

6. Ibid., pp. 5–7.

7. Ibid., p. 7.

8. Lois S. Peters and Herbert I. Fusfeld, "Current U.S. University-Industry Research Connections," in *University-Industry Research Relationships: Selected Studies,* National Science Foundation, Washington, D.C., 1982, p. 20.

9. Ibid., p. 49.

10. Ibid., p. 37.

11. Pat Hill Hubbard, *Planting the Engineering Seed Corn,* American Electronics Association, Palo Alto, 1981; Massachusetts High Technology Council, Proceedings of the Conference on Engineering Education, Boston, 1982.

12. Lois S. Peters and Herbert I. Fusfeld, "Current U.S. University-Industry Research Connections," in *University-Industry Research Relationships: Selected Studies,* National Science Foundation, Washington, D.C., 1982, p. 35.

13. Ibid., p. 34.

14. Kathryn Troy, *Annual Survey of Corporate Contributions: 1984 Edition,* The Conference Board, Report No. 848, New York, 1984; Dale Mann, *"All That Glitters: Public School/Private Sector Interaction in Twenty-Three U.S. Cities,"* unpublished report for the Exxon Education Foundation, Teachers College, Columbia University, 1984, pp. 18, 19.

15. Gerard G. Gold, "Toward Business–Higher Education Alliances," in

Business and Higher Education: Toward New Alliances, Gerard G. Gold (ed.), San Francisco: Jossey-Bass, 1981, pp. 9–27.

16. C. L. Aton, *An Advanced Technology Study for Post-Secondary Area Vocational-Technical Schools,* a report to the Georgia Office of Vocational Education, Georgia Department of Education, Georgia Institute of Technology, Atlanta, 1982. This study reviews vocational post-secondary programs across the country.

17. *San Jose Mercury News,* August 26, 1983, p. B1.

18. Judith K. Larsen and Carol Gill, *Changing Lifestyles in Silicon Valley,* Los Altos, Calif: Cognos Associates, 1983, p. 19.

19. Ann R. Nunez and Jill Frymier Russell, *As Others See Vocational Education, Book I: A Survey of the National Association of Manufacturers,* The National Center for Research in Vocational Education, Ohio State University, Columbus, Ohio, 1982.

20. National Science Foundation, *University-Industry Research Relationships,* Fourteenth Annual Report of the National Science Board, Washington, D.C., 1982, pp. 20–23.

21. Lois S. Peters and Herbert I. Fusfeld, "Current U.S. University-Industry Research Connections," in *University-Industry Research Relationships: Selected Studies,* National Science Foundation, Washington, D.C., 1982, p. 9.

22. James D. Marver and Carl V. Patton, "The Correlates of Consultation: American Academics in the 'Real World,'" *Higher Education,* 5 (1976): 319–335.

23. National Science Foundation, *University-Industry Research Relationships,* Fourteenth Annual Report of the National Science Board, Washington, D.C., 1982, p. 11.

24. Frank Darknell and Edith Darknell, "State College Science and Engineering Faculty: Collaborative Links with Private Business and Industry in California and Other States," in *University-Industry Research Relationships: Selected Studies,* National Science Foundation, Washington, D.C., 1982, pp. 163–192.

25. Henry Etzkowitz, "A Research University in Flux: University/Industry Interactions in Two Departments," State University of New York at Purchase, Report to the National Science Foundation, Washington, D.C., 1984; *Boston Globe,* November 7, 1983, p. 41.

26. National Science Foundation, *University-Industry Research Relationships,* Fourteenth Annual Report of the National Science Board, Washington, D.C., 1982, p. 29.

27. Ibid., p. 20.

28. Stanford University, "Fact Sheet" and press release on the Center for Integrated Systems, April 4, 1981.

29. Ibid.

30. Erich Bloch, James Meindl, and William Cromie, "University Cooperation in Microelectronics and Computers," in *University-Industry Research Relationships: Selected Studies*, National Science Foundation, Washington, D.C., 1982, p. 242.

31. Edward B. Roberts and Herbert A. Wainer, "New Enterprises on Route 128," *Science Journal*, December 1968; H. A. Wainer, "The Spin-Off of Technology from Government-Sponsored Research Laboratories: Lincoln Laboratory," unpublished master's thesis, MIT, Sloan School of Management, 1965; Dan A. Forseth, "The Role of Government-Sponsored Research Laboratories in the Generation of New Enterprises—A Comparative Analysis," unpublished master's thesis, MIT, Sloan School of Management, 1966.

32. National Science Foundation, *University-Industry Research Relationships*, Fourteenth Annual Report of the National Science Board, Washington, D.C., 1982, p. 20.

33. Ibid., p. 28.

34. Lois S. Peters and Herbert I. Fusfeld, "Current U.S. University-Industry Research Connections," in *University-Industry Research Relationships: Selected Studies*, National Science Foundation, Washington, D.C., 1982, p. 10.

35. Ibid., p. 63.

36. *Chronicle of Higher Education*, April 14, 1982, p. 9; Pat Hill Hubbard, *Planting the Engineering Seed Corn*, American Electronics Association, Palo Alto, 1981.

37. Pat Hill Hubbard, *Planting the Engineering Seed Corn*, American Electronics Association, Palo Alto, 1981; Massachusetts High Technology Council, Proceedings of the Conference on Engineering Education, Boston, 1982; "Industry Support for Education: A Principles and Practices Manual for Members of the Massachusetts High Technology Council, Second Edition," Boston, 1983.

38. *San Jose Mercury News*, April 29, 1982, p. B6.

39. Frank Darknell and Edith Darknell, "State College Science and Engineering Faculty: Collaborative Links with Private Business and Industry in California and Other States," in *University-Industry Research Relationships: Selected Studies*, National Science Foundation, Washington, D.C., 1982, pp. 175, 179.

40. Ibid., p. 175.

41. The Council for Financial Aid to Education, *Corporate Support of Education, 1983*, New York, 1984.

42. *San Jose Mercury News*, November 27, 1981, p. 1.

43. Peter Doeringer, Patricia Pannell, and Pankaj Tandon, "Market Influences on Higher Education: A Perspective for the 1980s," in *Business and Academia: Partners in New England's Economic*

Renewal, John C. Hoy and Melvin H. Bernstein (eds.), New England Board of Higher Education, Hanover, N.H.: University Press of New England, 1981, p. 73.

44. *Boston Globe*, February 15, 1982, p. 14, September 27, 1981, p. 26.
45. *Boston Phoenix*, April 28, 1981.
46. Ray Stata, "Appendix II. Perspectives of Two Business Leaders," in *Business and Academia: Partners in New England's Economic Renewal*, John C. Hoy and Melvin H. Bernstein (eds.), New England Board of Higher Education, Hanover, N.H.: University Press of New England, 1981.
47. *Chronicle of Higher Education*, December 7, 1983, p. 1, citing a study by the Southern Regional Education Board.
48. Lois S. Peters and Herbert I. Fusfeld, "Current U.S. University-Industry Research Connections," in *University-Industry Research Relationships: Selected Studies*, National Science Foundation, Washington, D.C., 1982, pp. 37–39, 112–116.
49. David F. Noble and Nancy E. Pfund, "Business Goes Back to College," *The Nation*, September 20, 1980, p. 251.
50. Lois S. Peters and Herbert I. Fusfeld, "Current U.S. University-Industry Research Connections," in *University-Industry Research Relationships: Selected Studies*, National Science Foundation, Washington, D.C., 1982, p. 119.

Chapter 8

1. *Education Week*, October 10, 1984, p. 8, citing a 1984 study by Science Service and Westinghouse Electric Corporation.
2. Gary Becker, *Human Capital*, New York: National Bureau of Economic Research, 1964; Thomas Juster (ed.), *Education, Income and Human Behavior*, New York: McGraw-Hill, 1975; Jacob Mincer, *Schooling, Earnings, and Experience*, New York: Columbia University Press, 1974. See Christopher Hurn, *The Limits and Possibilities of Schooling*, Boston: Allyn and Bacon, 1978, pp. 30–84, for a discussion of the functional and radical paradigms in the sociology of education. See also Randall Collins, "Functional and Conflict Theories of Educational Stratification," *American Sociological Review* 36 (1971): 1002–1019.
3. Samuel Bowles and Herbert Gintis, *Schooling in Capitalist America*, New York: Basic Books, 1976, pp. 235–239. This literature also includes, among others, Michael Katz, *The Irony of Early School Reform: Education and Innovation in Mid-Nineteenth Century Massachusetts*, Cambridge: Harvard University Press, 1968; Clarence J.

Karier, Joel Spring, and Paul C. Violas, *Roots of Crisis: American Education in the Twentieth Century*, Chicago: University of Illinois Press, 1973; Martin Carnoy and Henry Levin, *The Limits of Educational Reform*, New York: McKay, 1976; David Nasaw, *Schooled to Order: A Social History of Public Schooling in the United States*, New York: Oxford University Press, 1979. See Henry A. Giroux, "Theories of Reproduction and Resistance in the New Sociology of Education: A Critical Analysis," *Harvard Educational Review*, August 1983, pp. 257–293, for a comprehensive review of the radical paradigm.

4. Samuel Bowles and Herbert Gintis, *Schooling in Capitalist America*, New York: Basic Books, 1976; Michael W. Apple, *Ideology and Curriculum*, London and Boston: Routledge & Kegan Paul, 1979; David Hogan, "Making It in America: Work, Education, and Social Structure," in *Work, Youth, and Schooling: Historical Perspectives on Vocationalism in American Education*, Harvey Kantor and David B. Tyack (eds.), Stanford, Calif.: Stanford University Press, 1982, pp. 142–179.

5. Samuel Bowles and Herbert Gintis, *Schooling in Capitalist America*, New York: Basic Books, 1976, pp. 235–239.

6. Michael Timpane, *Corporations and Public Education in the Cities*, unpublished report for the Carnegie Corporation of New York, Teachers College, Columbia University, New York, 1982, pp. 23–24; Marsha Levine, "Barriers to Private Sector/Public School Collaboration: A Conceptual Framework," in *Barriers to Private Sector/Public School Collaboration*, Marsha Levine (ed.), Washington, D.C.: American Enterprise Institute and National Institute of Education, 1983, p. 10.

7. Samuel Bowles and Herbert Gintis, *Schooling in Capitalist America*, New York: Basic Books, 1976; Edward A. Krug, *The Shaping of the American High School, 1880–1920*, New York: Harper and Row, 1964; Marvin Lazerson, *Origins of the Urban School: Public Education in Massachusetts, 1870–1915*, Cambridge: Harvard University Press, 1971; David K. Cohen and Marvin Lazerson, "Education and the Corporate Order," *Socialist Revolution* 2 (March–April 1972): 47–72; Marvin Lazerson and W. Norton Grubb (eds.), *American Education and Vocationalism: A Documentary History, 1870–1970*, New York: Teachers College Press, 1974; Arthur G. Wirth, *Education in the Technological Society: The Vocational-Liberal Studies Controversy in the Early Twentieth Century*, Washington, D.C.: University Press of America, 1980; Harvey Kantor, "Vocationalism in American Education: The Economic and Political Context, 1880–1930," in *Work, Youth, and Schooling: Historical Perspectives on Vocationalism in American Education*, Harvey Kantor and David B.

Tyack (eds.), Stanford, Calif.: Stanford University Press, 1982; David Tyack and Elisabeth Hansot, *Managers of Virtue: Public School Leadership in America, 1820–1980*, New York: Basic Books, 1982.

8. Marvin Lazerson and W. Norton Grubb (eds.), *American Education and Vocationalism: A Documentary History 1870–1970*, New York: Teachers College, 1974, p. 90.

9. David Tyack and Elisabeth Hansot, *Managers of Virtue: Public School Leadership in America, 1820–1980*, New York: Basic Books, 1982, pp. 110 and 285.

10. Edward A. Krug, *The Shaping of the American High School, Volume 2, 1920–1941*, Madison, Wis.: University of Wisconsin Press, 1972, p. 206, quoting the December 1932 issue of *School Life*.

11. Michael W. Kirst, "Organizations in Shock and Overload: California's Public Schools 1970–1980," *Educational Evaluation and Policy Analysis* 1, 4 (July-August 1979): 27–30; *Education and Urban Society* 13, 2 (February 1981) (the entire issue of this journal is devoted to interest groups in education); David Tyack and Elisabeth Hansot, *Managers of Virtue: Public School Leadership in America, 1820–1980*, New York: Basic Books, 1982.

12. Michael W. Kirst, "Organizations in Shock and Overload: California's Public Schools 1970–1980," *Educational Evaluation and Policy Analysis* 1, 4 (July-August 1979): 27–30.

13. David Tyack and Elisabeth Hansot, *Managers of Virtue: Public School Leadership in America, 1820–1980*, New York: Basic Books, 1982, pp. 4, 8, 237–249.

14. Ibid., p. 6.

15. Ibid., p. 248.

16. Christopher J. Hurn, *The Limits and Possibilities of Schooling*, Boston: Allyn and Bacon, 1978, p. 217.

17. David Tyack and Elisabeth Hansot, *Managers of Virtue: Public School Leadership in America, 1820–1980*, New York: Basic Books, 1982, p. 249.

18. David Hogan, "Making It in America: Work, Education, and Social Structure," in *Work, Youth, and Schooling: Historical Perspectives on Vocationalism in American Education*, Harvey Kantor and David B. Tyack (eds.), Stanford, Calif.: Stanford University Press, 1982, pp. 142–179; Henry A. Giroux, "Theories of Reproduction and Resistance in the New Sociology of Education: A Critical Analysis," *Harvard Educational Review*, August 1983, pp. 257–293.

19. Alonzo A. Crim, "A Community of Believers," *Daedalus*, Fall 1981, pp. 145–162.

20. Susan Moore Johnson, "Merit Pay for Teachers: A Poor Prescription for Reform," *Harvard Educational Review*, May 1984, pp. 175–185;

Association for Supervision and Curriculum Development, "Incentives for Excellence in America's Schools," Alexandria, Va., 1985.

21. U.S. Bureau of the Census, "School Enrollment—Social and Economic Characteristics of Students: October, 1983," Washington, D.C.: U.S. Government Printing Office, 1984.

22. National Science Foundation, *University-Industry Research Relationships*, Fourteenth Annual Report of the National Science Board, Washington, D.C., 1982; National Science Board Commission on Precollege Education in Mathematics, Science and Technology, *Educating Americans for the 21st Century: A Report to the American People and the National Science Board*, National Science Foundation, Washington, D.C., 1983.

23. Gerald Holton, "*A Nation at Risk* Revisited," *Daedalus*, Fall 1984, p. 15.

Index